Pipe Welding Techniques

**IVAN H. GRIFFIN • EDWARD M. RODEN •
LARRY JEFFUS • CHARLES W. BRIGGS**

Pipe Welding Techniques

*This book is dedicated to two very special people,
my daughters Wendy and Amy.*

Photo courtesy of Linde Welding and Cutting Systems

Delmar Staff
   Administrative Editor: Mark W. Huth
   Production Editor: Carol A. Micheli

**For information, address Delmar Publishers Inc.
2 Computer Drive West, Box 15-015
Albany, New York 12212**

COPYRIGHT © 1985
BY DELMAR PUBLISHERS INC.

All rights reserved. Certain portions of this work copyright © 1965, 1972, and 1978. No part of this work covered by the copyright hereon may be reproduced or used in any form or by any means—graphic, electronic, or mechanical, including photocopying, recording, taping, or information storage and retrieval systems—without written permission of the publisher.

Printed in the United States of America
Published simultaneously in Canada
by Nelson Canada,
A division of The Thomson Corporation

10 9 8 7 6 5

Library of Congress Cataloging in Publication Data

Main entry under title:

Pipe welding techniques.

   Includes index.
   1. Pipe—Welding. I. Jeffus, Larry.
TS280.P49   1985   671.8'32   85-1513
ISBN 0-8273-2248-8

CONTENTS

Preface vi

SECTION 1 INTRODUCTION
Unit 1 Applications of Pipe 1
Unit 2 Codes and Standards for Pipe Welding............................. 7
Unit 3 Procedure Aids for Qualifying Tests 24

SECTION 2 PLATE WELDING
Unit 4 Root Pass 32
Unit 5 Single V-groove Vertical Welds 3G on Plate 45
Unit 6 Single V-groove Horizontal Welds 2G on Plate 53
Unit 7 Single V-groove Overhead Welds 4G on Plate 59

SECTION 3 PIPE WELDING
Unit 8 Beading.. 65
Unit 9 Fitting the Butt Joint 74
Unit 10 Welding Horizontal Pipe 1G and 5G Positions...................... 84
Unit 11 Welding Vertical Pipe 2G Position................................ 97
Unit 12 Welding 45-degree Inclined Pipe 6G and 6GR Positions 105

SECTION 4 SPECIAL APPLICATIONS
Unit 13 Gas Tungsten Arc Welding of Pipe................................. 109
Unit 14 Gas Metal Arc and Flux-cored Arc Welding of Pipe 119
Unit 15 Pipe Fitting... 126
Unit 16 90-degree Branch Connections..................................... 132
Unit 17 Lateral Pipe Connections... 137

Appendix I   Sizes and Types of Commercial Pipe......................... 142
Appendix II   Conversion Factors: U.S. Customary Units and SI Metric Units 143
Appendix III Welding Codes and Specifications........................... 145

Index.. 147

PREFACE

Pipe welding is one of the welding trades requiring the highest level of skill. A skilled pipe welder can expect to receive the admiration of other skilled tradespeople and a high rate of pay. The high skill levels required to perform quality pipe welds can be developed through the logical progression of the well-planned student lessons in *Pipe Welding Techniques.*

The 4th edition has been extensively revised and expanded to blend the content of earlier editions with the latest developments in pipe welding. Increased emphasis has been placed on actual welds made on pipe. The units on welding codes, weld testing, shielded-metal arc welding and pipe fitting have been revised to reflect the latest technology. New units have been added on gas-tungsten arc welding, gas-metal arc welding, and flux-cored arc welding. Many new drawings and photographs have been added throughout the book to better illustrate the units.

The units contain both technical information and detailed welding procedures. The early units help the beginning pipe welder make the transition from out-of-position plate welding to welding pipe. Once this transition is made, the student is given detailed instruction in welding pipe with a variety of processes in all positions. The major difference in welding flat plate and pipe is that the welding position on pipe changes constantly. The rate of change is faster on small-diameter pipe, so the beginner is encouraged to develop skill on large-diameter pipe first.

With this edition of *Pipe Welding Techniques* a new Instructor's Guide has been introduced. The Instructor's Guide includes answers to all text review questions, a comprehensive test, and masters from which visual aids can be produced.

This revision was prepared by Larry Jeffus. He is Welding Department Chairman at Eastfield College, Mesquite, Texas, a member of the AWS General Education Committee, Chairman of the North Texas section of the AWS, and author of the text, *Welding Principles and Applications*, by Delmar Publishers Inc.

ACKNOWLEDGMENTS

The author wishes to express appreciation to the following companies for the photographs that they supplied:

- Robvon Backing Ring Co.
- Hobart Brothers Co.
- Atlas Welding Accessories, Inc.
- G & W Products Inc.

Preface

- E. H. Wachs Company
- Vernon Tool Company
- H & M Pipe Beveling Machine Co., Inc.
- George A. Valdez, Photographer

NOTICE TO THE READER

Publisher does not warrant or guarantee, any of the products described herein or perform any independent analysis in connection with any of the product information contained herein. Publisher does not assume, and expressly disclaims, any obligation to obtain and include information other than that provided to it by the manufacturer.

The reader is expressly warned to consider and adopt all safety precautions that might be indicated by the activities described herein and to avoid all potential hazards. By following the instructions contained herein, the reader willingly assumes all risks in connection with such instructions.

The publisher makes no representations or warranties of any kind, including but not limited to, the warranties of fitness for particular purpose or merchantability, nor are any such representations implied with respect to the material set forth herein, and the publisher takes no responsibility with respect to such material. The publisher shall not be liable for any special, consequential or exemplary damages resulting, in whole or in part, from the readers' use of, or reliance upon, this material.

Section 1

INTRODUCTION

The pipe welder has always been considered one of the most highly skilled and most respected of all welders. This level of admiration is well-justified because the job they do is indeed difficult. In almost every situation their welds are very critically inspected so that even the slightest defect will be located and repaired. The weld is usually around a fixed pipe which may be in a hard-to-get-to location. The combination of all these factors has led to the pipe welder being one of the highest paid welding jobs.

Unit 1

APPLICATIONS OF PIPE

USES OF PIPE

More than one-tenth of the entire United States' yearly steel production is used in the manufacture of pipe: this is millions of tons of iron, steel, and alloy steel pipe. In addition, the amount of nonferrous pipe such as aluminum, copper, and nickel being fabricated is increasing rapidly. This enormous production of pipe requires a workforce in the millions to install and maintain piping systems.

How is all this pipe used? It serves as a container and conveyor for a wide variety of products. Pipelines deliver gas, petroleum, and water cross-country. Specially constructed pipe filled with insulating oil under pressure carries high-voltage power cables. Slurries of powdered coal and water are transported through pipelines. Iron ore is delivered through pipelines from the mines to the processor.

Industrial piping is another large application of pipe. Refineries, chemical plants, atomic energy plants, and power and processing plants in which pressures are high and heat and corrosion are severe, all utilize large quantities of pipe.

Other uses of pipe include plumbing and welded pipe structures. To be strong and rigid, pipe structures must be designed properly and welded skillfully.

Figure 1-1 An impressive pipe installation marks this new aromatics plant in Puerto Rico. The island is emerging as a petrochemical center which has the raw material capacity for over 800 product possibilities.

Pipe Welding Techniques

METHODS OF JOINING PIPE

Pipe may be joined together to form a system by the use of threaded fittings, clamp-on fittings, flanges, brazing, soldering, or welding, Figure 1-2.

Some of the major advantages of welded-pipe connections are:

- strength. The thickness of the pipe at the connection is not reduced by cutting threads so that the completed joint is as strong as the surrounding pipe, Figure 1-3.
- less maintenance required. The joint is less likely to leak even after long periods of time in service.
- smooth flow. There is no change in internal size that can cause turbulance in the material as it flows through the fitting, Figure 1-4.
- lighter weight. There is no heavy fitting required at each joint which can add considerable weight to the system.

REQUIREMENTS OF WELDED-PIPE SYSTEMS

Many piping systems are subjected to extremely high pressures and often to high temperatures as well. Cross-country gas pipelines must operate under high pressures to supply adequate amounts of this highly volatile fuel. Any gas leaking through improperly welded pipe joints is a fire or explosion hazard. Piping in the petroleum and chemical industries may be subjected to high temperature and corrosive action as well as high pressure. This often requires the use of special alloys and special welding procedures. Fossil fuel plants—those using

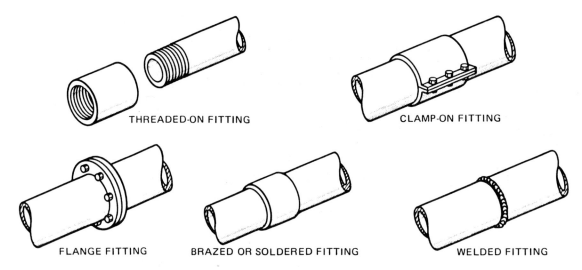

Figure 1-2 Pipe fittings

Applications of Pipe

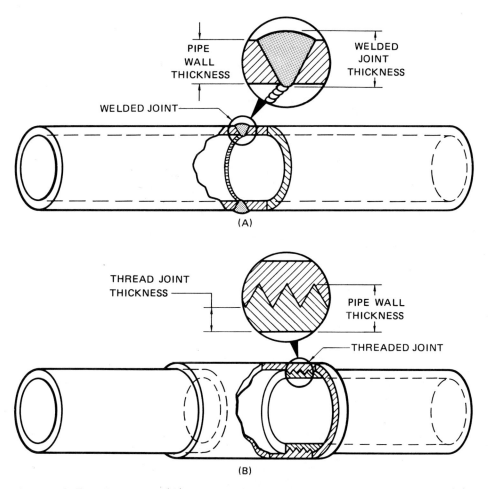

Figure 1-3 The welded joint (A) is thicker than the original pipe; the threaded joint (B) is thinner than the original pipe. (From Jeffus & Johnson, *Welding Principles and Applications.* Copyright 1984 by Delmar Publishers Inc.)

coal, oil, or natural gas for the generation of electricity—typically operate at about 2,400 pounds of pressure per square inch (psi) at 1,000 degrees Fahrenheit. Nuclear electric plants operate at 950 psi and 540 degrees Fahrenheit. Their piping systems are constructed of welded alloy pipe designed to withstand these pressures and temperatures.

On the opposite end of the temperature scale cryogenic piping systems are welded and installed to transmit and contain such elements as liquid oxygen and liquid hydrogen which require extremely high pressures and low temperatures in their production and storage.

The variations in pressure, temperature, and corrosion have required pipe manufacturers to produce pipe and tubing in many different sizes, wall thicknesses, and alloys.

Pipe Welding Techniques

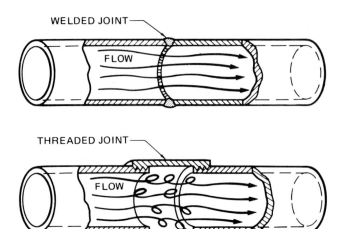

Figure 1-4 The flow along a welded pipe is less turbulent than that in a threaded pipe. (From Jeffus & Johnson, *Welding Principles and Applications.* Copyright 1984 by Delmar Publishers Inc.)

Figure 1-5 Field welding gas pipe. Note that a helper assists the welder.

Applications of Pipe

WELDING STANDARDS

In order to ensure high standards of uniformity, safety, and excellence, most pipe is designed, installed, and tested under one or more of the various codes or standards set up by such societies and associations as:

- The American Society of Mechanical Engineers (ASME)
- The American National Standards Institute (ANSI)
- The American Petroleum Institute (API)
- The American Welding Society (AWS)
- The American Society for Testing and Materials (ASTM)
- The American Iron and Steel Institute (AISI)
- The American Institute of Steel Construction (AISC)

Some insurance companies, manufacturers, the military, and other governmental agencies set up their own standards and codes to govern the fabrication and welding of piping systems and their components. These codes or standards are strict and often specify the type of test the welder must pass, the frequency of retesting of the welder, and the preweld and post-

Figure 1-6 Welding a dry well assembly at the Nine Mile Point Nuclear Station at Lake Ontario.

Pipe Welding Techniques

weld heat treatment and procedures. Most manufacturers of piping components subject welders to a very rigid qualifying test with provision for retesting every six months.

REVIEW

Select the letter preceding the best answer.

1. In order to ensure high-quality welds, most pipe is designed, installed, and tested
    a. according to textbooks.
    b. by state officials.
    c. by visual examinations.
    d. under one or more of the various codes or standards.

2. Pipe welding codes are set up by
    a. associations and societies.
    b. state governments.
    c. insurance companies, manufacturers, and the military.
    d. All of the above.

3. Large-scale, high-pressure welded piping systems are most frequently used for
    a. fossil-fueled electric power generating systems.
    b. drainage systems in manufacturing plants.
    c. transportation of powdered coal.
    d. nuclear electric generating systems.

4. Stresses caused by high pressures in welded piping systems are most severe in
    a. urban water distribution systems.
    b. nuclear electric generating plants.
    c. electric generating plants fueled by coal.
    d. alloy pipe which conducts corrosive liquids.

5. What are piping systems joined with clamp-on fittings best used for?
    a. Low pressure
    b. Corrosive chemicals
    c. Water supply
    d. Cryogenics

Unit 2

CODES AND STANDARDS FOR PIPE WELDING

Most pipe welding is carried out under one of the various codes or standards that govern the conditions that must be met if a welded piping system is to be approved. In actual practice, most of the material published in the various codes is in the form of data and specifications published primarily for use by engineering departments. The testing of welds according to these standards has two purposes: (1) to qualify the procedure necessary to produce welds of acceptable quality, and (2) to qualify welders to work on the piping system. These and other organizations that publish codes and standards dealing with piping and pressure vessels are:

- The American Society of Mechanical Engineers (ASME)
  United Engineering Center
  345 East 47th Street
  New York, New York 10017

- The American National Standards Institute (ANSI)
  1430 Broadway
  New York, New York 10018

- The American Petroleum Institute (API)
  1801 K Street, NW
  Washington, D.C. 20006

CODE WELDING

Code welding refers to the construction of weldments by a certified welder according to the specifications of a particular code. The code selected for use is decided by the manufacturer, insurance company, contractor, or other governing agency. The most widely accepted and commonly used codes are ASME and API. The student of pipe welding should understand that there are two separate areas to be considered in all welding codes: each procedure used to construct a weldment must be tested and approved before it is considered qualified and welders must pass a separate performance test for each procedure they wish to perform.

Procedure Qualifications

Procedure qualifications for a particular job must first be written in the form of a Welding Procedure Specification. Figure 2-1 shows the suggested form used for a job to comply

Pipe Welding Techniques

WELDING PROCEDURE SPECIFICATION (WPS)

Company Name _____
Welding Procedure Specification No. _____ Date _____ Supporting PQR No. _____
                Revisions _____   _____

Welding Process(es) _____ Type(s) _____

JOINTS
Groove Design _____
Backing: Yes _____ No _____
Backing Material (Type) _____
Other _____

BASE METALS
P No. _____ to P. No. _____
Thickness Range _____
Pipe Dia. Range _____
Other _____

FILLER METALS
F No. _____ Other _____
A No. _____ Other _____
Spec No. (SFA) _____
AWS No. (Class) _____
Size of Electrode _____
Size of Filler _____
Electrode-Flux (Class) _____
Consumable Insert _____
Other _____

POSITION
Position of Groove _____
Welding Progression _____
Other _____

PREHEAT
Preheat Temp. _____
Interpass Temp. _____
Preheat Maintenance _____
Other _____

POSTWELD HEAT TREATMENT
Temperature _____
Time Range _____
Other _____

GAS
Shielding Gas(es) _____
Percent Composition (mixtures) _____
Flow Rate _____
Gas Backing _____
Trailing Shielding Gas Composition _____
Other _____

ELECTRICAL CHARACTERISTICS
Current AC or DC _____ Polarity _____
Amps (Range) _____ Volts (Range) _____
Other _____

TECHNIQUE
String or Weave Bead _____
Orifice or Gas Cup Size _____
Initial & Interpass Cleaning (Brushing, Grinding, etc) _____
Method of Back Gouging _____
Oscillation _____
Contact Tube to Work Distance _____
Multiple or Single Pass (per side) _____
Multiple or Single Electrodes _____
Travel Speed (Range) _____
Other _____

This form may be obtained from the Order Dept., ASME, 345 E. 47th St., New York, N.Y. 10017

Figure 2-1

with the ASME *Boiler Pressure Vessel Code,* Section IX. This specification lists the exact procedure to be followed to construct the necessary weldment. Such specifications include the:

- welding process to be used
- material to be welded
- joint design
- position
- technique to be used
- type and size of electrodes
- contour and degree of bevel of the edges
- direction of welding
- amperage
- type of current (AC, DCSP, DCRP)
- rate of travel
- number of passes
- backing or chill rings when necessary
- temperature of the material in the vicinity of the joint during welding
- postweld heat treatment when necessary

Upon completion of the test specimen by a qualified welder, it is tested according to the code being used. If the test weldment passes all of the required tests, a Procedure Qualification Record must be completed and kept on file. This form, Figure 2-2, contains all the information regarding the construction of the weld and the results of the tests performed on the weldment.

Welder Performance Qualifications

Once a manufacturer, insurance company, contractor, or other governing agency has approved a welding procedure, a welder qualified for that particular procedure is needed to complete the operation. Each welder must qualify to construct the necessary weldment by passing the required performance test. In most cases the tests may be completed on the job through the use of a test pipe according to the code. Tests are conducted by the employer or governing agency and the results are recorded on a special form, Figure 2-3, and kept on file.

Tests are arranged in specific positions as shown in Figure 2-4. Each of the test positions is assigned a label. To be qualified to weld these positions on the job, the welder must pass the test required for each position.

Pipe Welding Techniques

PROCEDURE QUALIFICATION RECORD (PQR)

Company Name _____
Procedure Qualification Record No. _____ Date _____
WPS No. _____
Welding Process(es) _____
Types (Manual, Automatic, Semi-Auto.) _____

JOINTS

Groove Design Used

BASE METALS
Material Spec. _____
Type or Grade _____
P No. _____ to P No. _____
Thickness _____
Diameter _____
Other _____
_____
_____
_____

POSTWELD HEAT TREATMENT
Temperature _____
Time _____
Other _____

GAS
Type of Gas or Gases _____
Composition of Gas Mixture _____
Other _____

FILLER METALS
Weld Metal Analysis A No. _____
Size of Electrode _____
Filler Metal F No. _____
SFA Specification _____
AWS Classification _____
Other _____
_____

ELECTRICAL CHARACTERISTICS
Current _____
Polarity _____
Amps. _____ Volts _____
Other _____

POSITION
Position of Groove _____
Weld Progression (Uphill, Downhill) _____
Other _____
_____

TECHNIQUE
Travel Speed _____
String or Weave Bead _____
Oscillation _____
Multipass or Single Pass (per side) _____
Single or Multiple Electrodes _____
Other _____
_____

PREHEAT
Preheat Temp. _____
Interpass Temp. _____
Other _____

This form may be obtained from The Order Dept., ASME, 345 E. 47th St., New York, N.Y. 10017

Figure 2-2

Tensile Test

Specimen No.	Width	Thickness	Area	Ultimate Total Load lb.	Ultimate Unit Stress psi	Character of Failure & Location

Guided Bend Tests

Type and Figure No.	Result

Toughness Tests

Specimen No.	Notch Location	Notch Type	Test Temp.	Impact Values	Lateral Exp. % Shear	Lateral Exp. Mils	Drop Weight Break	Drop Weight No Break

Fillet Weld Test

Result – Satisfactory: Yes _____ No _____ Penetration into Parent Metal: Yes _____ No _____
Macro – Results _____

Other Tests

Type of Test _____
Deposit Analysis _____
Other _____

Welder's Name _____ Clock No. _____ Stamp No. _____
Tests conducted by: _____ Laboratory Test No. _____
We certify that the statements in this record are correct and that the test welds were prepared, welded and tested in accordance with the requirements of Section IX of the ASME Code.

Manufacturer _____
Date _____ By _____
(Detail of record of tests are illustrative only and may be modified to conform to the type and number of tests required by the Code.)

Figure 2-2 (continued)

Pipe Welding Techniques

MANUFACTURER'S RECORD OF WELDER OR WELDING OPERATOR QUALIFICATION TESTS

Welder Name _____ Check No. _____ Stamp No. _____
Welding Process _____ Type _____
In accordance with Welding Procedure Specification (WPS) _____
Backing _____
Material — Spec. _____ to _____ of P No. _____ to P No. _____
     Thickness _____ Dia. _____
Filler Metal (QW-404) Spec. No. _____ Class No. _____ F No. _____
     Other _____
Position (1G, 4F, 6G, etc.) _____
Gas — Type _____ % Composition _____
Electrical Characteristics — Current _____ Polarity _____
Weld Progression _____
Other _____

For Information Only

Filler Metal Diameter and Trade Name _____
Submerged Arc Flux Trade Name _____
Gas Metal Arc Welding Shield Gas Trade Name _____

Guided Bend Test Results

Type and Fig. No.	Results

Radiographic Test Results
For alternative qualification of groove welds by radiography

Radiographic Results: _____

Fillet Weld Test Results

Fracture Test (Describe the location, nature and size of any crack or tearing of the specimen) _____
_____
Length and Per Cent of Defects _____ inches _____ % _____
Macro Test — Fusion _____
Appearance — Fillet Size (leg) _____ in. X _____ in. Convexity _____ in. or Concavity _____ in.

Test Conducted by _____ Laboratory — Test No. _____
We certify that the statements in this record are correct and that the test welds were prepared, welded and tested in accordance with the requirements of Sections IX of the ASME Code.

          Organization _____
Date _____ By _____

(Detail of record of tests are illustrative only and may be modified to conform to the type and number of tests required by the Code.)
NOTE: Any essential variables in addition to those above shall be recorded.
    This form may be obtained from the Order Dept., ASME, 345 E. 47th St., New York, N.Y. 10017

Figure 2-3

Codes and Standards for Pipe Welding

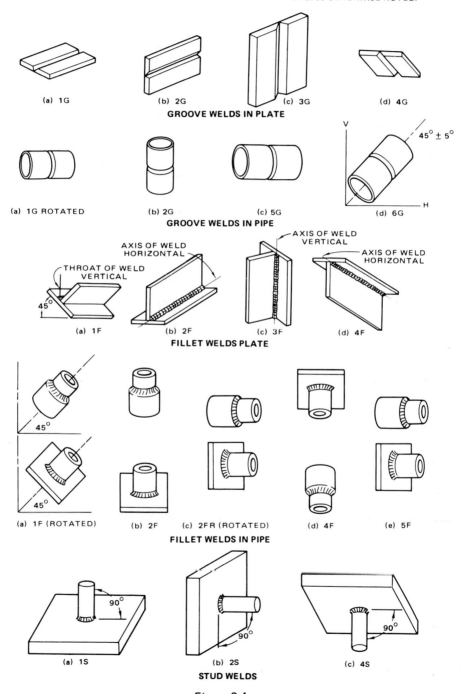

Figure 2-4

Pipe Welding Techniques

The ASME recommends that the gas-tungsten arc welding (GTAW) process be used for the root pass on sections for nuclear power piping. The pipe welding student must possess a basic competency in GTAW (TIG) welding as well as shielded-metal arc welding (SMAW).

The direction of welding is usually not considered foremost, and the choice is generally the welder's, except in the case of vertical welds. Procedure specifications determine whether these welds should proceed from the bottom upward (vertical-up) or from the top downward (vertical-down).

Qualification in certain difficult positions may enable the welder to perform similar, less difficult welds. The ASME Code states:

QW-303 Limits of Qualified Positions (See QW-461)

QW-303.1 **Groove Welds—General.** Qualification in the 2G, or 3G, or 4G position shall also qualify for the 1G position. Qualification in the 5G position shall qualify for the 1G, 3G, and 4G positions. Qualification in both the 2G and 5G positions, or in the 6G position, shall qualify for all positions.

QW-303.2 **Groove Welds—Positions 1G and 2G Only—Single Welds with Backing.** Qualification on single-welded plate with backing shall also qualify for single-welded pipe with backing 2 7/8 inch outside diameter and over, and qualification on any diameter single-welded pipe with backing, shall also qualify for single-welded plate, with backing, in positions 1G and 2G only.

QW-303.3 **Groove Welds—Position 1G and 2G Only—Single Welds without Backing.** Qualification on single-welded plate without backing shall also qualify for single-welded pipe with or without backing 2 7/8 inch outside diameter and over, and qualification on any diameter single-welded pipe without backing, shall also qualify for single-welded plate with or without backing, in positions 1G and 2G only.

QW-303.4 **Groove Welds—Positions 1G and 2G Only—Double Welds.** Qualification on double-welded plate shall also qualify for double-welded pipe 2 7/8 inch outside diameter and over, and qualification on any diameter double-welded pipe shall also qualify for double-welded plate, in positions 1G and 2G only.

QW-303.5 **Other Positions.** For all other positions, qualification on pipe shall qualify for plate, but not vice versa, except that qualification on plate shall qualify for pipe over 24 inches in diameter.

QW303.6 **Fillet Welds—General.** Qualification in the 2F, 3F, 4F, or 5F position shall also qualify for the 1F position. Qualification in the 3F, 4F, or 5F position shall also qualify for the 2F position. Qualification in both 3F and 4F positions for plate, or the 5F position for pipe, shall qualify for all F positions.

Qualification in the 2G position shall also qualify for the 1F and 2F positions. Qualification in the 5G or 6G position shall qualify for all positions of fillet welds.

Welders who pass the required tests for groove welds shall also be qualified to make fillet welds of any size on base metals in all thicknesses and pipe diameters, within the limits of the welding variables in QW-350 (the code). Welders who pass the tests for fillet

Codes and Standards for Pipe Welding

welds shall be qualified to make fillet welds only in all thicknesses of material, sizes of fillet welds, and diameters of pipe or tube 2 7/8 inch outside diameter and over, for use within the other applicable essential variables. Welders who make fillet welds on pipe or tube less than 2 7/8 inch outside diameter must pass the pipe fillet-weld test per QW-452.4 (the code) or the required tests for groove welds.

QW-303.7 Special Positions. A fabricator who does production welding in a special orientation may make the tests for performance qualification in this specific orientation. Such qualifications are valid only for the positions actually tested, except that an angular deviation of plus or minus 15 degrees is permitted in the inclination of the weld axis and the rotation of the weld face, as defined in QW-461.1 (the code).

TESTING METHODS

Testing methods for pipe welding are classified as nondestructive and mechanical testing. In general, nondestructive methods are used to test piping systems and pressure vessels, whereas destructive (mechanical) methods are used for procedure tests and welder performance-qualifying tests.

Visual Test (VT). All welds must first pass a visual inspection test before they are tested further. During the visual inspection the weld's location, size, and other surface conditions are evaluated for acceptability. The inspector may use a number of devices to aid in the test. Gauges such as fit-up, alignment, depth, spacing, and weld size are often used, Figure 2-5.

Penetrant Testing (PT). The use of one of the two types of liquid for the penetrant test aids in locating and identifying those surface discontinuities that may be too small or not otherwise easily located. Either a colored dye or a flourescent dye (which shows up under black light), is applied to a clean, dry surface and allowed to seep into the flaws. Then the excessive dye is removed and a developing powder is applied. The powder acts like a blotter to draw the dye out of any flaws, making locating them easier, Figure 2-6.

Ultrasonic Testing (UT). Because of the low cost, higher speed and high degree of accuracy, ultrasonic testing is used on almost all pressure vessels. This inspection method uses a high-frequency sound wave (similar to sonar) to penetrate the metal and locate any discontinuities, Figure 2-7. Using a properly calibrated machine a skilled inspector can located and identify most internal problems in a weldment.

Proof Testing (PRT). The most common proof test still in use today is the *Hydrostatic Test*. In this test the pressure vessel is filled with water and then a pump is used to raise the pressure to a predetermined point. The test pressure is usually 50% or 100% of the expected working limits of the vessel.

Most forms of proof testing are no longer used because it is believed that the tests may actually cause problems that can reduce the product's reliability.

Pipe Welding Techniques

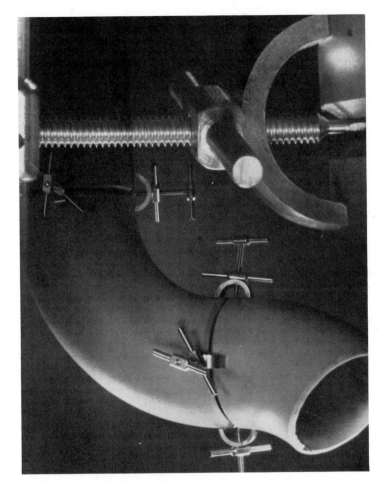

Figure 2-5 Clamps for holding pipes in place for welding. (Photo courtesy of G & W Products Inc.

Radiographic Testing (RT). When exceptionally high-quality welds are required, the radiographic method of inspection is favored. It is used not only for piping systems, but also for testing welding procedures and qualifying operators.

Inspection is accomplished with X rays or radioactive isotopes, usually cobalt 60 or iridium 192. X-ray equipment is usually limited to shop radiography.

Destructive Methods

The common types of destructive physical testing are carried out by using *test specimens* cut from test plates or test pipes. The test plate or pipe is the larger piece of metal

Codes and Standards for Pipe Welding

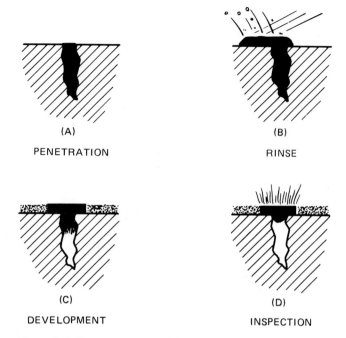

Figure 2-6 Penetrant testing (From Jeffus & Johnson, *Welding Principles and Applications.* Copyright 1984 by Delmar Publishers Inc.)

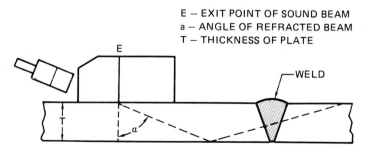

Figure 2-7 Ultrasonic testing (From Jeffus & Johnson, *Welding Principles and Applications.* Copyright 1984 by Delmar Publishers Inc.)

Pipe Welding Techniques

containing the welded joint. This joint is welded in accordance with the requirements and specifications of the code being used.

The two methods of destructive testing which will be discussed in this unit are the guided-bend test and the reduced-section tension test. The guided-bend test is commonly used for welder qualification, while both the reduced-section tension test and the guided-bend test are commonly used for establishing welding procedures.

The Preparation of Specimens. The prepare specimens for testing, they must be cut to certain dimensions. Refer to Figure 2-8 for the preparation of specimens from plate and pipe. Figure 2-9 shows the location of specimen removal from welded test plate. Individual specimens should be stamped with steel letters and number dies on the face side of the plates. All the backing strips and weld reinforcements should be removed flush with the base metal. This may be done by grinding, shaping, or milling. All tool marks or grinding marks should be across the weld, Figure 2-10. All four corners of the specimens taken from plate should be filed to a 1/16-inch radius.

The Guided-bend Test. The guided-bend test is the most common test used to qualify welders. In this test, the specimens are bent in the jig at the face and root of the weld through 180 degrees and examined for signs of failure. In the case of 3/8-inch groove welds, one specimen is bent with the root of the weld on the convex side. For the fillet soundness test, two specimens are bent with the root of the weld on the convex side.

The convex side of each completed test specimen is examined for signs of lack of fusion, Figure 2-11 and 2-12. It is also inspected for gas pockets, slag inclusions, and cracks. Any of

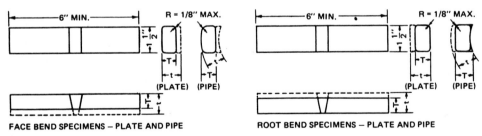

t, in.	T, in.	
	Metals in QW-422.35 and QW-422.23	All Ferrous Metals
1/16-1/8	t	t
1/8-3/8	1/8	t
>3/8	1/8	3/8

Note 1: Weld reinforcement and backing strip or backing ring, if any, shall be removed flush with the surface of the specimen. If a recessed ring is used, this surface of the specimen may be machined to a depth not exceeding the depth of the recess to remove the ring, except that in such cases the thickness of the finished specimen shall be that specified above. Do not flame-cut nonferrous material.

Note 2: If the pipe being tested has a 3-inch outside diameter or less, the width of the bend specimen may be 3/4 inch, measured around the outside surface. If the pipe being tested is less than 2-inch pipe size (2.375 in. outside diameter), the width of the bend specimens may be that obtained by cutting the pipe into quarter sections, less an allowance for saw cuts or machine cutting.

Figure 2-8 Face and root bend specimens: plate and pipe

Codes and Standards for Pipe Welding

Figure 2-9 Location of specimen removal from welded test plates

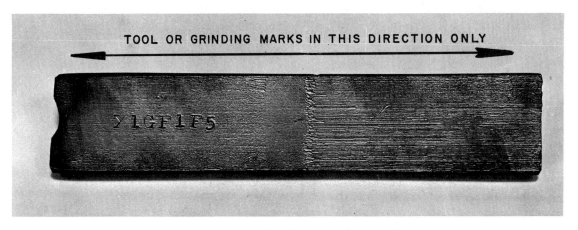

Figure 2-10 Direction of tool marks

these faults, which appear after bending and which measure 1/8 inch or more in any direction, is cause for failure in the test.

When the guided-bend test is used to qualify welders, and any of the welds in any position fails, the welder can qualify by immediately making two test plates for each unsuccessful position. All test specimens must pass to qualify.

In this text, the student is asked to test the practice welds by means of the guided-bend jig. If there is a need to fabricate a jig, Figure 2-13 gives the necessary details.

Pipe Welding Techniques

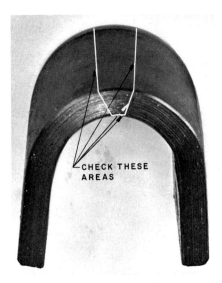

Figure 2-11 Completed test

Figure 2-12 Weld defects after bending

The Reduced-section Tension Test. The reduced-section tension test is generally used for testing welding procedures. The test is made in a tensile-testing machine capable of pulling the specimen apart. A dial indicates the tensile strength of the weld in pounds per square inch. The weld passes the test if the specimen breaks in the base metal outside the welded zone. If the fracture occurs in the weld zone, the psi reading must be higher than that of the base metal.

Codes and Standards for Pipe Welding

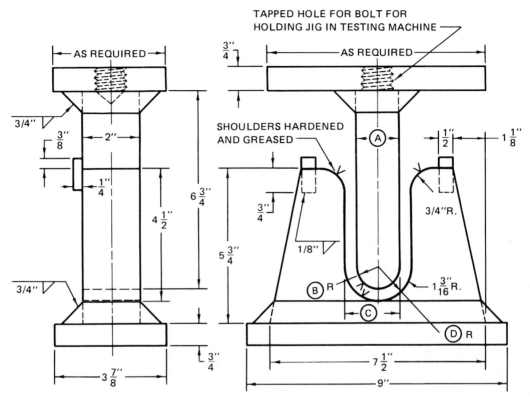

THICKNESS OF SPECIMENS, IN.	A IN.	B IN.	C IN.	D IN.	MATERIAL
3/8	1 1/2	3/4	2 3/8	1 3/16	All
t	4t	2t	6t + 1/8	3t + 1/16	Others

Figure 2-13 Guided-bend jig

Figure 2-9 shows the location for the removal of the various specimens from plate. Figure 2-14 shows the location for the removal of the various specimens from pipe. Figure 2-15 illustrates the preparation of the reduced-section tension specimen.

Pipe Welding Techniques

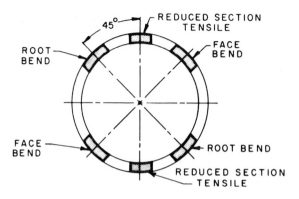

Figure 2-14 Location of specimen removal from welded pipe

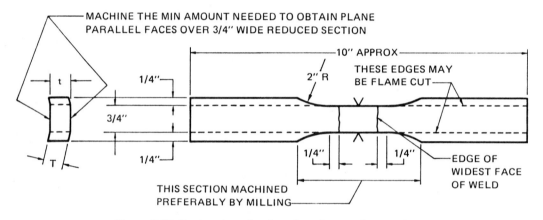

Figure 2-15 Preparation of reduced-section tension specimen

REVIEW

Select the letter preceding the best answer.

1. Which organization publishes the most commonly used code for boiler and pressure vessel welding?
   a. American Welding Society
   b. American Society of Mechanical Engineers
   c. American Petroleum Institute
   d. American National Standards Institute

Codes and Standards for Pipe Welding

2. The ASME recommends that the root pass on sections for nuclear power piping be deposited by the
   a. oxyacetylene process.
   b. shielded metal-arc welding process.
   c. metallic inert-gas process.
   d. tungsten inert-gas process.

3. According to the ASME code, which test position qualifies the welder for all pipe groove welds?
   a. 1F b. 4F c. 4G d. 6G

4. After bending a specimen, which of the following is not considered to be a test failure?
   a. Lack of fusion
   b. Small surface defects
   c. Slag inclusions
   d. Pockets caused by gas

5. When conducting a guided-bend test,
   a. use a low-carbon steel specimen.
   b. use a high-carbon steel specimen.
   c. bend the specimen through 180 degrees.
   d. bend the specimen until a crack develops.

Unit 3

PROCEDURE AIDS FOR QUALIFYING TESTS

A number of variables can affect a welder's success in passing a qualifying test. This unit will investigate the errors that frequently cause failure of the test specimens. Also, those areas will be pointed out where a thorough knowledge of the test procedures will help the welder to produce specimens of acceptable quality.

MATERIALS

Material for practicing pipe-welding techniques should be low-carbon (mild) steel plate which possesses good ductility for making the guided-bend test. Before the material is used for test plates a sample should first be guided-bend tested to see if it has the required ductility, Figure 3-1 A and B.

Figure 3-1A A preweld base metal specimen of acceptable quality

Procedure Aids for Qualifying Tests

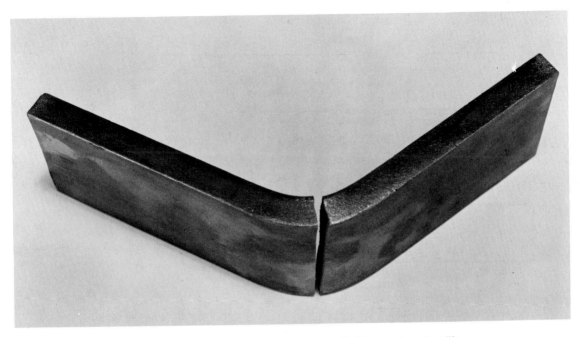

Figure 3-1B High-strength material that failed due to low ductility

The test plates or pipe may be sheared, flame-cut, or machined to the proper size and bevel of the code being used. Figure 3-2 is a cross-sectional view of the test plate edge preparation and a setup for the ASME code.

The electrodes should be the standard group F3, E6010 or E6011 classification, or low-hydrogen E7018 type. Transmission piping is generally welded with E6010 electrodes. Industrial power piping, such as nuclear power piping, is welded with E7018 electrodes. Low-hydrogen electrodes are specified on some jobs because they produce good results in out-of-position welding; tend to reduce cracking and porosity in alloy steels, especially those containing sulfur; clean up easily to reduce inner pass and post-weld cleanup time; and the resulting beads display high strength and ductility.

Electrode specifications relating to size and current vary according to the type of joint, the position of the test plate, and the direction of welding. Generally, electrodes that are 3/32 inch (2.4 mm) and 1/8 inch (3 mm) in diameter are used for most pipe welding.

This text will emphasize the use of both of the F3 group electrodes (E6010 and E6011), since they are the standard electrodes used for most transmission pipe welding. The SMAW low-hydrogen (E7018) electrode and the GTAW process will be covered but not as extensively.

PREPARATION, POSITIONING, AND WELDING

For passing the guided-bend test, there is an advantage if the test plates can be cut so that the welding is done across the direction of rolling. Bar stock and plate are rolled the

Pipe Welding Techniques

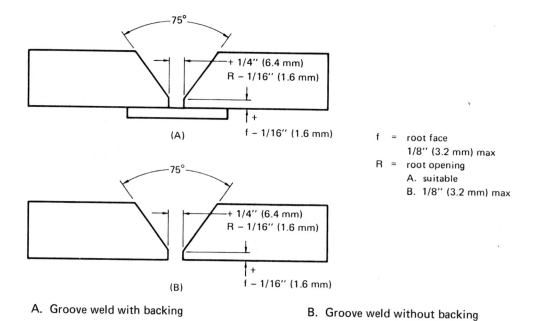

A. Groove weld with backing   B. Groove weld without backing

Figure 3-2 Joint preparation for guided-bend testing

long way. For example, 3/8-inch x 4-foot x 10-foot plate is rolled in the 10-foot direction, **Figure 3-3**.

Mill scale should always be removed from the test plates and backing strips. Any oxides left on the test plates can combine with the weld deposit to produce welds of lower strength and ductility.

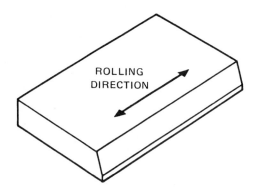

Figure 3-3 Beveling the edge in the same direction the plate was rolled will help you pass the guided-bend test

Procedure Aids for Qualifying Tests

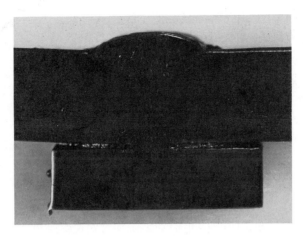

Figure 3-4 Macroetched section of multiple-pass weld

Personal comfort should be the guide when welding out of the flat position. Plates positioned too low or too high in relation to eye level make it difficult for the welder to observe the molten puddle.

The number of passes will vary with the following:

- size of the electrode
- direction of welding
- position of the weld
- type of electrode

The welding student must remember that a weld consisting of a number of thin beads is metallurgically more desirable than a weld of one thick bead. This characteristic is shown in Figure 3-4. Note that each succeeding bead tends to refine the grain structure of the preceding bead. In order for this grain reduction to occur most efficiently, the plate must not be welded on when it is at the wrong temperature. It must be above 70°F (21°C) and below 300°F (135°C) for most thicknesses and types of low-carbon steel, Figure 3-5.

MACROETCHING

The process which exposes the grain pattern in this manner is called *macroetching*. This type of weld examination may be conducted by the welding student with little difficulty. The following procedure explains the steps required to macroetch a weld specimen.

Pipe Welding Techniques

Figure 3-5A Thermocouple leads are spot welded to the pipe, at least 1 inch (25 mm) from the joint.

Figure 3-5B The thermocouple leads are covered with a high-temperature insulation.

PROCEDURE

1. Cut a cross-section specimen from the weld.
2. Grind and polish the specimen to a fine finish.
   Note: The final finish can be produced on a surface grinder, if one is available. If not, place a sheet of #500 abrasive cloth on plate glass and stroke the specimen, face down, on this cloth until all scratches are removed.
3. Swab the polished surface with a saturated solution of ammonium persulphate, (9 to 1 solution of water and ammonium persulphate). Observe the etching process until the grain structure is clearly visible. Wash the specimen in warm water and dry with compressed air. Do not rub the surface with cloth or paper towels. This tends to make the etch less distinct.

Procedure Aids for Qualifying Tests

Figure 3-5C Themocouple lead ends are clamped under a bolt and nut that is welded to the pipe, at least 1 inch (25 mm) from the joint. (Photographs courtesy of George A. Valdez, photographer)

4. To preserve the specimen, spray the *dry* surface with a clear, quick-drying material, such as clear Krylon® or one of the clear lacquers. Some lacquers last indefinitely without discoloring.

Test plates should be allowed to cool to 200°F (93°C) before any additional beads are deposited. The temperature of the plates can be checked by sprinkling a few drops of water on them. If the water boils, the temperature is at least 212°F (100°C). Higher temperature specifications are checked with appropriate temperature-indicating crayons.

REVIEW

Select the letter preceding the best answer.

1. A specimen is tested in the guided-bend test by
    a. breaking in a vise.
    b. twisting with a wrench.
    c. bending in a jig.
    d. bending over a pipe.
2. The different grain structure between the weld deposit and the base metal can be determined by
    a. a face-bend test.
    b. a root-bend test.
    c. a hardness test.
    d. an etching test.

Pipe Welding Techniques

3. A root-bend test is used to check the amount of weld
    - a. ductility.
    - b. elongation.
    - c. hardness.
    - d. penetration.

4. The ductility of steel for test specimens is determined by the
    - a. etching test.
    - b. tensile test.
    - c. hardness test.
    - d. guided-bend test.

5. When welding test plates from which test specimens will be cut,
    - a. weld in the direction of the length of the stock.
    - b. remove oxides and scale from the test plates before welding.
    - c. additional beads may be deposited if the temperature of the plate is over 300°F.
    - d. use only nickel steel test plates.

Section 2
PLATE WELDING

Most welding codes and standards require that welders first pass a welding test on plate if they are going to be welding on large diameter systems. The welding tests, are usually given in the 3G; vertical, 2G; horizontal, and 4G; overhead positions. Some shops will allow the welders to be certified if they pass the 3G and 4G test.

In this section you will be both improving welding skills that you already have and developing the skills to weld on large-diameter pipe and pressure vessels.

Unit 4

ROOT PASS

In this unit you will learn the welding technique required to make a root weld in plate. The root pass is used to establish the depth of penetration. This is the basis of a quality weld and is essential for the weld to pass the root-bend test.

A welded joint can be classified by whether it has a open root or a backing strip, Figure 4-1. The open root is the most common type of joint in both plate and pipe welding. For that reason most of the welds practiced in this book will be open root-type joints.

MATERIALS

- Steel plates, 3/8 inch (10 mm) x 1 inch (25 mm) x 6 inches (152 mm)*
- 3/32 inch (2.3 mm) or 1/8 inch (3 mm) diameter E6010, E6011 and E7018 electrodes.

VERTICAL—UP 3G

Procedure E6010 and E6011 Open Root

1. Tack weld the plates together as shown in Figure 4-2.
2. Position the assembled plates in the vertical position at a comfortable height.
3. Hold the electrode at a slightly upward angle (15 degrees) and strike an arc at the bottom of one of the joints.

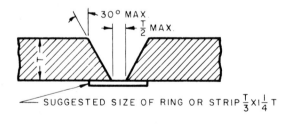

BUTT JOINT FOR WELDER QUALIFICATION
(FOR PLATE OR PIPE "T" IN. THICK)

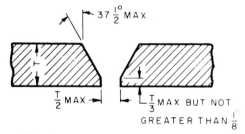

ALTERNATIVE BUTT JOINT FOR WELDER QUALIFICATION
(FOR PLATE OR PIPE "T" IN. THICK)

Figure 4-1 Edge preparation

*1/8 inch (3 mm) thick steel strips can be used for practicing the root pass. Using the thinner metal for practice will save time because it does not require beveling.

Root Pass

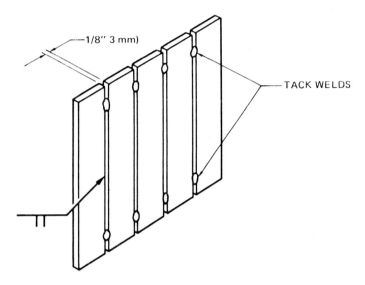

Figure 4-2 Vertical root weld plates

Note: Try not to strike the arc out of the joint on the surrounding plate surface. This is a skill you must develop because most codes consider arc strikes outside of the groove to be defects.

4. Adjust both your movement and the machine current setting so that you can maintain a key hole just ahead of the weld bead, Figure 4-3.
   Note: If the root opening and current are correct then a slow, straight upward electrode movement can be used. But if the root opening is either too close, too wide, or varies, and the current is too high you must use a stepping electrode movement to make a good root pass, Figure 4-4.
   Note: Maintaining a key hole is a good way to insure that the root of the joint is being fused properly.
5. Clean and examine the first bead. Look closely at the root surface to see that it is uniform and does not have either excessive reinforcement or underfill. Although having a uniform root face is important, it is not as important as having a good root surface. Most problems with the root face are easily corrected by grinding or the use of a filler pass; this is not always possible with the root surface, Figure 4-5.
6. Repeat these welds using the E6010 and E6011 electrodes with both the straight and the stepping movements until you have mastered them.

Pipe Welding Techniques

Procedure E7018 Open Root

1. Tack weld the plates together as shown in Figure 4-2.
2. Position the assembled plates in the vertical position at a comfortable height.
3. Hold the electrode at a slightly upward angle (15 degrees) and strike an arc at the bottom of one of the joints.
4. Adjust both your movement and the machine current setting so that you can maintain a key hole just ahead of the weld bead.
   Note: The movement of the E7018 electrode should be less than 2 1/2 times the diameter of the electrode.
5. Clean and examine the first bead. Look closely at the root surface to see that it is uniform and does not have either excessive reinforcement or underfill.
6. Repeat the weld using the E7018 electrode until you have mastered it.

Procedure E6010 and E6011 with a Backing Strip

1. Tack weld the plates together as shown in Figure 4-6.
2. Position the assembled plates in the vertical position at a comfortable height.
3. Hold the electrode at a slightly upward angle (15 degrees) and strike an arc at the bottom of one of the joints.
4. Adjust both your movement and the machine current setting so that you can maintain complete root fusion.

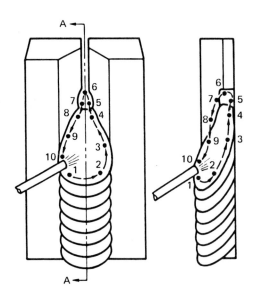

Figure 4-3 Electrode movement to open and use a key hole (From Jeffus & Johnson, *Welding Principles and Applications.* Copyright 1984 by Delmar Publishers Inc.)

Root Pass

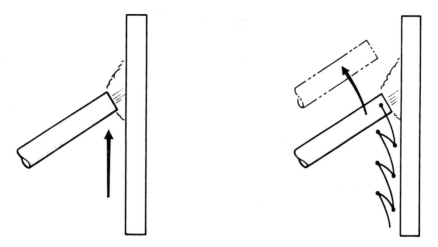

Figure 4-4 Electrode movement

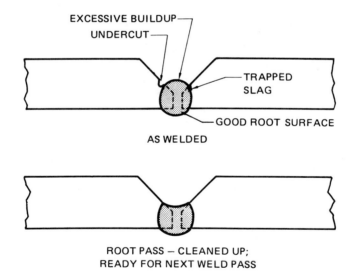

Figure 4-5 Grinding back the root pass to ensure a sound second pass. (From Jeffus & Johnson, *Welding Principles and Applications.* Copyright 1984 by Delmar Publishers Inc.)

Note: Watch the two inside corners of the root to see that there is good fusion so that slag is not trapped along the edges, Figure 4-7.
5. Clean and examine the first bead. Look closely at the root face to see that it is uniform and does not have slag trapped along the side.
6. Repeat these welds using the E6010 and E6011 electrodes with both the straight and the stepping movements until you have mastered them.

Pipe Welding Techniques

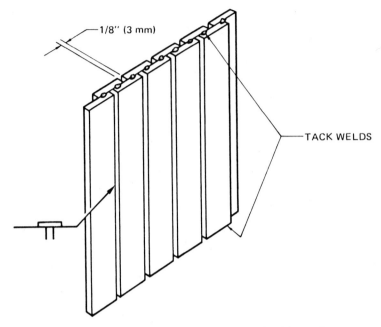

Figure 4-6 Vertical root weld plate with backing strip

Procedure E7018 with a Backing Strip

1. Tack weld the plates together as shown in Figure 4-6.
2. Position the assembled plates in the vertical position at a comfortable height.
3. Hold the electrode at a slightly upward angle (15 degrees) and strike an arc at the bottom of one of the joints.
4. Adjust both your movement and the machine current setting so that you can maintain complete root fusion.
   Note: A slightly higher current setting may be required to help prevent slag from being trapped along the edges.

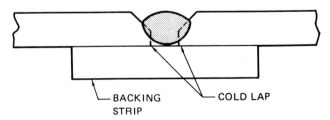

Figure 4-7 Incomplete root fusion. (From Jeffus & Johnson, *Welding Principles and Applications.* Copyright 1984 by Delmar Publishers Inc.)

Root Pass

5. Clean and examine the first bead. Look closely at the root face to see that it is uniform and does not have slag trapped along the side.
6. Repeat this weld using the E7018 electrode until you have it mastered.

VERTICAL-DOWN 3G

The vertical-down root pass is a technique that can be used reliably when making welds *only* if the joint fitup is correct. This technique makes good-looking welds but it is more difficult to make sound welds than with the vertical-up technique. The smooth uniform root face is often misleading because, unless care is taken, slag will run ahead of the molten weld pool and be trapped under the weld bead, Figure 4-8.

Vertical-down welding is faster than vertical-up; once the skills are developed they can be used to make as good a quality of weld as any other procedure.

Procedure E6010 and E6011 Open Root

1. Tack weld the plates together as shown in Figure 4-2.
2. Position the assembled plates in the vertical position at a comfortable height.

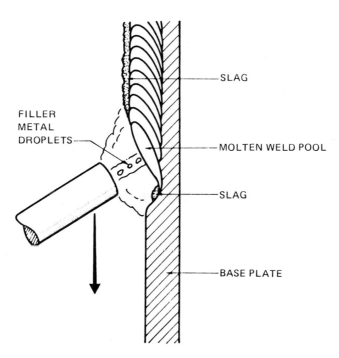

Figure 4-8 The arc force moves the molten filler across to the weld pool.

Pipe Welding Techniques

3. Hold the electrode at a slightly upward angle (15 degrees) and strike an arc at the top of one of the joints.
4. Adjust both your movement and the machine current setting so that you can maintain a key hole just ahead of the weld bead.
   Note: If the root opening and current are correct then a slow, straight downward electrode movement can be used. If the root opening is either a little close or wide and the current is not too high a stepping electrode movement is better to keep the key hole from getting too large, Figure 4-9.
5. Clean and examine the first bead. Look closely at the root surface to see that it is uniform and does not have either excessive reinforcement or underfill.
6. Repeat these welds using the E6010 and E6011 electrodes until you master them.

Procedure E6010 and E6011 with a Backing Strip

1. Tack weld the plates together as shown in Figure 4-6.
2. Position the assembled plates in the vertical position at a comfortable height.
3. Hold the electrode at a slightly upward angle (15 degrees) and strike an arc at the top of one of the joints.

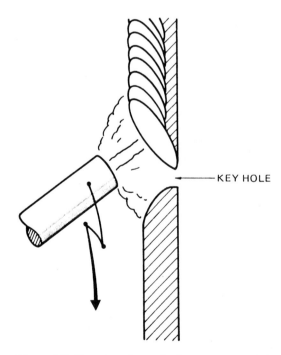

Figure 4-9 Stepping electrode down to control the size of the key hole

Root Pass

4. Adjust both your movement and the machine current setting so that you can maintain complete root fusion.
   Note: Watch the two inside corners of the root to see that there is good fusion so that slag is not trapped along the edges, Figure 4-7.
5. Clean and examine the first bead. Look closely at the root face to see that it is uniform and does not have slag trapped along the side.
6. Repeat these welds using the E6010 and E6011 electrodes until you master them.

HORIZONTAL 2G

Procedure E6010 and E6011 Open Root

1. Tack weld the plates together as shown in Figure 4-2.
2. Position the assembled plates in the horizontal position at a comfortable height.
3. Hold the electrode at a slightly upward angle (15 degrees) and strike an arc at one end of the joint.
4. Adjust both your movement and the machine current setting so that you can maintain a key hole just ahead of the weld bead.
5. Clean and examine the first bead. Look closely at the root surface to see that it is uniform and does not have either excessive reinforcement or underfill.
6. Repeat these welds using the E6010 and E6011 electrodes until you master them.

Procedure E6010 and E6011 with a Backing Strip

1. Tack weld the plates together as shown in Figure 4-6.
2. Position the assembled plates in the horizontal position at a comfortable height.
3. Hold the electrode at a slightly upward angle (15 degrees) and strike an arc at one end of the joint.
4. Adjust both your movement and the machine current setting so that you can maintain complete root fusion.
5. Clean and examine the first bead. Look closely at the root face to see that it is uniform and does not have slag trapped along the side.
6. Repeat these welds using the E6010 and E6011 electrodes until you master them.

Procedure E7018 with a Backing Strip

1. Tack weld the plates together as shown in Figure 4-6.
2. Position the assembled plates in the horizontal position at a comfortable height.
3. Hold the electrode at a slightly (15 degrees) upward angle and strike an arc at one end of the joint.
4. Adjust both your movement and the machine current setting so that you can maintain complete root fusion.
5. Clean and examine the first bead. Look closely at the root face to see that it is uniform and does not have either excessive reinforcement or underfill.

Pipe Welding Techniques

6. Repeat the weld using the E7018 electrode until you have mastered it.

Procedure E7018 with a Backing Strip

1. Tack weld the plates together as shown in Figure 4-6.
2. Position the assembled plates in the horizontal position at a comfortable height.
3. Hold the electrode at a slightly upward angle (15 degrees) and strike an arc at one end of the joint.
4. Adjust both your movement and the machine current setting so that you can maintain complete root fusion.
5. Clean and examine the first bead. Look closely at the root face to see that it is uniform and does not have slag trapped along the side.
6. Repeat the weld using the E7018 electrode until you have it mastered.

OVERHEAD 4G

Note: When welding in the overhead position it is important to wear proper protective clothing, Figure 4-10.

Figure 4-10 Full leather jacket (From Jeffus & Johnson, *Welding Principles and Applications.* Copyright 1984 by Delmar Publishers Inc.)

Root Pass

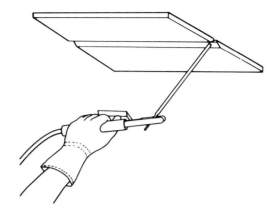

Figure 4-11 Holding your hand forward of the welding arc will not only help to keep the weld in view, it will keep your hand out of the main stream of weld spatter.

Procedure E6010 and E6011 Open Root

1. Tack weld the plates together as shown in Figure 4-2.
2. Position the assembled plates in the overhead position at a comfortable height.
3. Hold the electrode at a slightly forward angle and strike an arc at one end of the joint, Figure 4-11.
4. Adjust both your movement and the machine current setting so that you can maintain a key hole just ahead of the weld bead.
5. Clean and examine the first bead. Look closely at the root surface to see that it is uniform and does not have either excessive reinforcement or underfill.
6. Repeat these welds using the E6010 and E6011 electrodes until you master them.

Procedure E6010 and E6011 with a Backing Strip

1. Tack weld the plates together as shown in Figure 4-6.
2. Position the assembled plates in the overhead position at a comfortable height.
3. Hold the electrode at a slightly forward angle and strike an arc at one end of the joint.
4. Adjust both your movement and the machine current setting so that you can maintain complete root fusion.
5. Clean and examine the first bead. Look closely at the root face to see that it is uniform and does not have slag trapped along the side.
6. Repeat these welds using the E6010 and E6011 electrodes until you master them.

Procedure E7018 with a Backing Strip

1. Tack weld the plates together as shown in Figure 4-6.
2. Position the assembled plates in the overhead position at a comfortable height.

Pipe Welding Techniques

3. Hold the electrode at a slightly forward angle and strike an arc at one end of the joint.
4. Adjust both your movement and the machine current setting so that you can maintain complete root fusion.
5. Clean and examine the first bead. Look closely at the root face to see that it is uniform and does not have either excessive reinforcement or underfill.
6. Repeat the weld using the E7018 electrode until you have mastered it.

Procedure E7018 with a Backing Strip

1. Tack weld the plates together as shown in Figure 4-6.
2. Position the assembled plates in the overhead position at a comfortable height.
3. Hold the electrode at a slightly forward angle and strike an arc at one end of the joint.
4. Adjust both your movement and the machine current setting so that you can maintain complete root fusion.
5. Clean and examine the first bead. Look closely at the root face to see that it is uniform and does not have slag trapped along the side.
6. Repeat the weld using the E7018 electrode until you have it mastered.

REVIEW

Select the letter preceding the best answer.

1. To correct excessive root reinforcement you should
   a. lower the current setting.
   b. use a longer arc length.
   c. change from an E6010 to an E6011 electrode.
   d. use a shorter stepping motion.

2. What is the angle that the electrode should be held at for a vertical-up root pass?
   a. 25° up
   b. 10° down
   c. 15° up
   d. Vertical

3. What is the key hole used for?
   a. To see the root surface
   b. To help insure good root fusion
   c. To let the sparks fall away from you
   d. So that the molten weld metal has a place to flow through

4. Why is the condition of the root surface so important?
   a. It can not be easily repaired.
   b. If it is not smooth the weld will not pass the bend test.

c. It is the basis of a good weld.
d. All of the above.

5. What part of a root pass should you watch if a backing strip is used?
   a. The trailing edge
   b. The leading edge in side corners
   c. The root face
   d. The root surface

6. Why is it easy to have slag trapped in the vertical-down weld?
   a. The welding current is often too low.
   b. There is not enough heat to burn it out.
   c. Slag can run ahead of the molten weld pool.
   d. The welder is traveling too fast.

7. What is the advantage of vertical-down?
   a. It's faster than vertical-up.
   b. There is less skill required to make a good weld.
   c. The weld is always better.
   d. Less post-weld cleanup is required.

8. Why is undercutting along the root face not a major problem?
   a. The root pass is not as critical as other passes.
   b. The filler pass will cover it up.
   c. There is very little stress on the root.
   d. All of the above.

9. When would you use vertical-down welding?
   a. If the root opening is wider than normal
   b. When there is little importance in the weld
   c. If the root face is to be removed
   d. If the fit-up is correct

10. What can be done if the key hole is getting too large?
    a. Lower the current setting.
    b. Use a stepping motion.
    c. Use a larger electrode.
    d. Both a and b.

11. What protective clothing must be worn for overhead welding?
    a. Gauntlet leather gloves
    b. Leather jacket
    c. Welder's cap
    d. All of the above.

Pipe Welding Techniques

12. Why is it important to learn to make good open root welds?
    a. Because most pipe joints are of that type.
    b. The root of the weld must be good to pass the bend test.
    c. Most pipe is held in a fixed position.
    d. Large diameter pipe tests are given on plate.

Unit 5

SINGLE V-GROOVE VERTICAL WELDS 3G ON PLATE

Vertical-up and vertical-down welding each have advantages in specific applications. A comparison of the two methods shows that vertical-up welding allows greater penetration of the base metal and that irregularities in the size of the weld and the joint are more easily corrected during welding. Although the slower travel speed in vertical-up welding allows the welder to see the work more clearly, it is more difficult to master the technique of proper electrode manipulation.

The greatest advantage of vertical-down welding is its speed, which lowers the cost of the process. Although it is easier for the welder to master the technique of bead application, careless technique is more apt to cause slag inclusions and gas pockets. The direction in which a vertical weld is deposited is usually specified in the procedure qualifications, Figure 5-1.

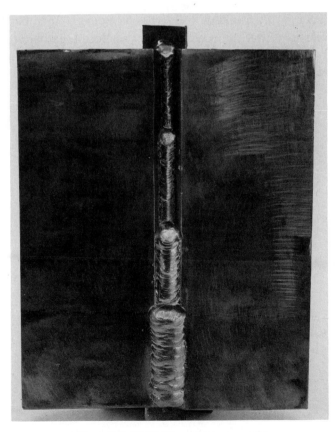

Figure 5-1 Vertical-up V-groove weld

Pipe Welding Techniques

MATERIALS

- 3/8 inch x 3 inch x 6 inch steel plate
- 3/8 inch x 1 inch x 7 inch steel backing strip
- DC welding machine
- E6010, E6011 and E7018 electrodes

VERTICAL-UP

Procedure with a Backing Strip

1. Tack the plates as shown in Figure 5-2. Make sure that all the precautions outlined in Unit 3 for removing oxides and dirt are followed.
2. Position the plates vertically at eye level.
3. Start the first bead at the bottom of the joint, using a motion and cadence similar to that used when welding the first bead in Unit 4. As this weld proceeds upward, watch the molten pool.
   Note: When practicing this weld with E7018 electrodes, remember that the arc length is kept short and that the beads are kept as close together as possible.
4. Cool, clean, and inspect this root pass for slag inclusions, voids, and lack of fusion. These are the causes of failure when guided face-bend and root-bend tests are made. All three

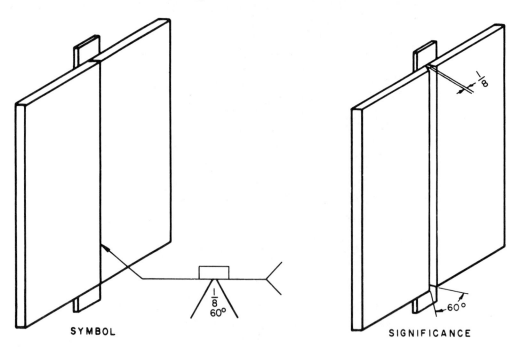

Figure 5-2 Vertical beveled butt joint with a backing strip

Single V-groove Vertical Welds 3G on Plate

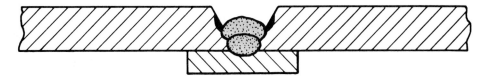

Figure 5-3 Undercut second bead with slag in undercut

of the faults are usually caused by too low a current setting, too short a pause on the down cycle of the weave, or trying to reach out too far with each succeeding cycle. The wrong arc length or electrode angle also contribute to a faulty bead.

5. Make the second pass. Keep the molten pool as large as possible while keeping the weld free from excessive sag and undercut. Observe the molten pool closely when making this weave and bring the arc down on the opposite side fast enough so that the molten pools join. Allowing the pool to solidify completely on one side before establishing the pool on the opposite side increases the probability of slag inclusions or voids.
6. Cool, clean, and inspect the weld. Look for sag, voids, slag inclusions, and undercut. Undercutting at this point makes it difficult to produce a sound third bead because it is difficult to remove all the slag in the undercut area. Figure 5-3 shows the cross section of an undercut second bead.
7. Make the third pass. Weave the electrode and arc just enough so that the center of the electrode is at the edge of the bevel on each side of the joint. This will provide for full fusion without making this pass unduly wide.
8. Cool and clean this pass. Examine it for sag, voids, slag inclusions, and undercut.
9. Examine the edges very critically for any evidence of undercutting, however slight.

    Note: Remember that the procedure for machining test specimens requires that all the weld reinforcing is removed but none of the base metal. One of the prime causes of failure in the face-bend tests is areas of undercut which show up as slight notches in the specimen after machining. They increase the probability of failure during the guided-bend test.

10. Prepare and weld more test plates until consistently sound joints can be produced with all beads of such quality that they can pass the most rigid visual inspection.
11. Straighten the cooled test plates. Lay them out for cutting and mark them for bending.
12. Cut the plates into test specimens and have them machined or ground as indicated in the machining instructions, Unit 3.
13. File the corners of the specimens to a 1/16-inch radius.
14. Test the specimens by the guided-bend test.
15. Inspect the bent specimens as indicated in Unit 3. Failure at this point calls for additional practice on this type of joint.

Procedure Open Root

1. Tack the plates together as shown in Figure 5-4. Make sure that all of the precautions outlined in Unit 3 for removing oxides and dirt are followed.

Pipe Welding Techniques

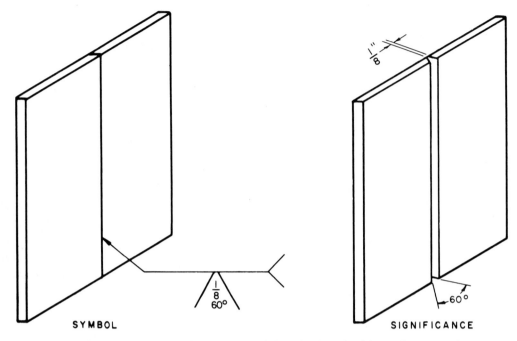

Figure 5-4 Vertical bevel butt joint without a backing strip

2. Position the plates vertically at eye level.
3. Start at the bottom using either the straight or stepping motion learned in Unit 4.
4. Cool, clean, and inspect before continuing with the remaining weld passes as you did on the weld with the backing strip.

VERTICAL-DOWN

Procedure with a Backing Strip E6010 and E6011

1. Tack weld plates and backing strip into position, Figure 5-2.
2. Position the steel plates at eye level.
3. With the electrode pointed upward at 20 to 25 degrees, establish an arc at the top of the plate and weld down without weaving, using a medium-short arc and a higher-than-normal rate of travel.
4. Observe the molten pool closely. Correct the tendency of the pool to run ahead of the arc by using a shorter arc, a greater electrode angle, a higher rate of travel, or a combination of all three.
    a. Cool and clean the first bead and lay down the second bead with a straight back-and-forth motion, using a short arc at all times. Compare the length of the pause at the sides of the weave with the cadence used to produce a similar bead by vertical-up welding. If the molten pool tends to run ahead of the arc, use slightly less current and a shorter arc.

Single V-groove Vertical Welds 3G on Plate

b. Examine this bead for signs of slag inclusions or holes and correct for these faults when making additional beads.
c. Make the third bead like the second one but use a wider side-to-side weave. Produce this weld with the same cadence, arc length, and rate of advance used for the second bead.
d. If inspection of this bead shows it to be too shallow, use a fourth bead to complete the joint.
   Note: Vertical-down welds do not tend to build up as thick a weld at each pass as vertical-up welds. An attempt to complete the joint with this third pass may result in an inferior weld whereby too large a pool is maintained due to too slow a rate of travel.
e. Inspect the completed weld for any holes or slag inclusions.
f. Check the edges of the weld for a slightly concave appearance with some excess build-up, Figure 5-5. If this occurs, correct it in subsequent beads by varying the speed of the weave motion. Pause for a shorter time at the edges and move more slowly from side to side. Figure 5-6 indicates the correct contour of a well-made vertical-down weld.
g. Once consistently good welds can be produced in this position, cut test specimens and test them as previously instructed.

Procedure Without a Backing Strip E6010 and E6011

1. Prepare and tack weld plates as shown in Figure 5-4, for practice in welding beveled joints without a backing strip.
2. Start the arc at the top of the assembly with the electrode pointed upward about 30 degrees. Once the arc has been established, allow the electrode coating to contact the

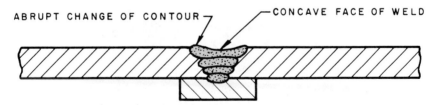

Figure 5-5 Cross section of vertical-down bead with improper shape and contour

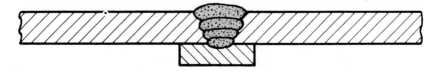

Figure 5-6 Correct contour for vertical-down weld

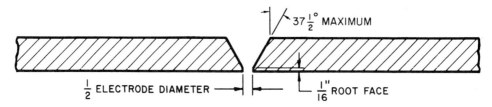

Figure 5-7 Plates beveled and set up for vertical welds—no backing strip

work and proceed downward at a rapid rate. It should be possible to produce a sound bead without burn-through or holes if the current setting, electrode angle, and rate of travel are adjusted to the gap indicated in Figure 5-7.
3. Clean and examine both sides of the joint. Pay particular attention to the back of the plate. It should look like a small weld without slag inclusions or holes, Figure 5-8.
4. Continue to make this type of weld until sound welds can be produced each time. If possible, machine or grind the back of the plate flush so that any faults will show up clearly during inspection.
5. Complete all the joints by making the filler passes and the cap pass. At this point the welder should have gained enough skill and experience to try making the entire joint in not more than three passes.
6. When the weld appears sound, cut test specimens and make the guided-bend test. At the stage, failure generally occurs in the root tests and indicates the need for more practice. Failures are caused by too wide a root opening, a faulty current setting, or the wrong electrode angle. Compare the results with those from the first procedure.

Figure 5-8 Reverse side of vertical-down root pass

Single V-groove Vertical Welds 3G on Plate

REVIEW

Select the letter preceding the best answer.

1. When manipulating a low-hydrogen electrode,
   a. the arc should be held close.
   b. a long arc is required.
   c. a shipping motion of the electrode should be used.
   d. weave beads should be used.

2. Low-hydrogen electrodes differ from the E6010 electrodes in that they
   a. are acceptable for use in the horizontal position.
   b. require less manipulation.
   c. increase the susceptibility to bead cracking.
   d. are available only in the E7010 classification.

3. What is the chief disadvantage of vertical-down welding over vertical-up welding?
   a. Penetration tends to be too deep.
   b. Vertical-down welding requires more time.
   c. There is a tendency to build up a thicker weld.
   d. Faults may be washed over, thus hiding holes or slag.

4. What are the relative merits of vertical-up welding compared to vertical-down welding as far as test results are concerned?
   a. Both types of joints test out equally well.
   b. Vertical-up welding produces superior test results.
   c. Vertical-down welding produces superior test results.
   d. Testing of vertical-down welds exposes below-the-surface defects of previous passes.

5. In vertical-down welding on beveled plate, the molten pool tends to run ahead of the arc. How can this be corrected?
   a. Reduce the current below 150 amps.
   b. Use a higher rate of travel.
   c. Increase the arc length.
   d. Use a smaller electrode angle.

6. What is the most common cause of failure in root-bend tests?
   a. Too high a current setting
   b. Too long a pause on the down cycle of the weave
   c. Lack of fusion and penetration
   d. Too high a travel speed

7. When running the first bead at the bottom of the beveled butt joint, keep the pool as molten as possible to
   a. avoid fusion at the root of the weld.
   b. prevent undercutting.

Pipe Welding Techniques

    c. provide effective penetration into the backing strip.
    d. maintain the electrode in a horizontal position.

Unit 6

SINGLE V-GROOVE HORIZONTAL WELDS 2G ON PLATE

Procedure with a Backing Strip E6010 and E6011

1. Clean all surfaces as suggested in Unit 3.
2. Set up and tack the plates and backing strip as indicated in Figure 6-1.
3. Position the plates vertically, with the groove in a horizontal position, at approximately eye level.
4. Run the root pass with the electrode pointed slightly upward, Figure 6-2.
5. Clean this bead and examine it for signs of undercut along the top of the plate and sag along the bottom of the bead.
   Note: Undercutting is due to too long an arc or too high a rate of travel. Sag is due to too high a current setting or too low a rate of travel.
6. Make the second pass by manipulating the electrode and arc so that the molten puddle has an elliptical shape with the long axis about 20 to 25 degrees from the vertical. Pause at the top of the weave for a fast 1, 2, 3 count, using a very short arc. Then proceed forward and downward at an angle of about 20 degrees. Do not pause at the bottom of the weave but proceed rapidly to the top of the stroke and repeat the cycle. Figure 6-3 indicates the shape of the crater, the angle of the electrode, and the manipulation of the arc.
   Note: The welder should realize that smoothness of arc manipulation is essential to produce uniform beads. These are much easier to clean than irregular beads. All beads must be thoroughly cleaned of all slag before any additional beads are deposited.

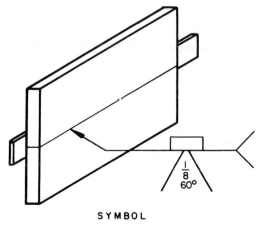
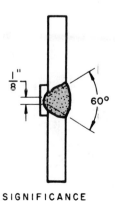

SYMBOL · SIGNIFICANCE

Figure 6-1 Horizontal welds on beveled plate

Pipe Welding Techniques

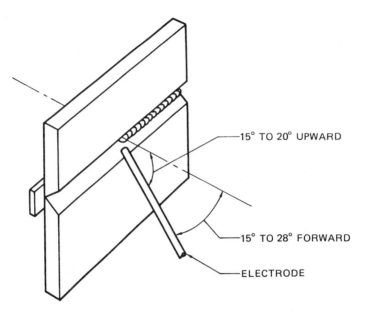

Figure 6-2 First pass on horizontal weld

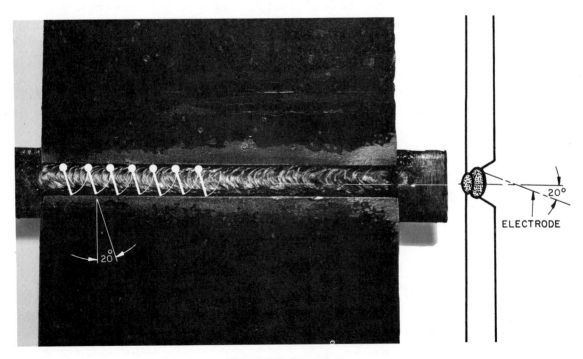

Figure 6-3 Second pass

Single V-groove Horizontal Welds 2G on Plate

7. Clean and inspect the second bead as in step 5.
8. Deposit the final pass in much the same manner as the second pass.
    Note: Beginners tend to make this pass too wide and usually too heavy or thick. In actual practice, the center of the electrode should coincide with the edge of the bevel at both the top and bottom of the weave. Figure 6-4 indicates the electrode angle, the angle of the weave pattern, and the areas in which pauses and rapid manipulation are necessary.
9. Clean and inspect the final bead, checking for the faults indicated in step 5. Be especially critical of this bead. It is no more important than the preceding beads, but this is the bead that the customer, inspector, or foreman sees. Any imperfection or irregularity in this bead, however small, will surely leave doubts in their minds as to the soundness of all underlying beads.
10. After good welds can be produced consistently, cut test specimens and test them as instructed in Units 2 and 3.
11. The backing strip and weld reinforcement should be removed by machining or grinding. Grinding marks should run across the weld.

Procedure with a Backing Strip E7018

1. Duplicate the first five steps of the prior procedure (E6010 and E6011).
    Note: No manipulation is required in this position when using the low-hydrogen electrode. The completed weld is made with a series of small beads, Figure 6-5. It is possible and sometimes desirable to use this same sequence with the E6010 and E6011 electrodes.

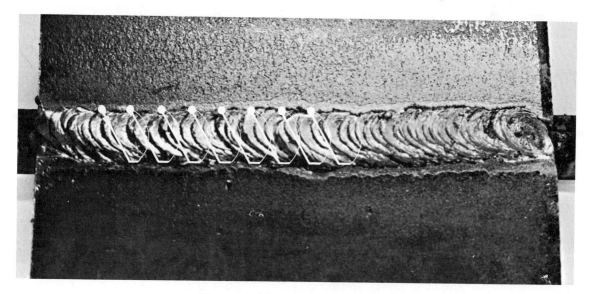

Figure 6-4 Weave bead for third pass

Pipe Welding Techniques

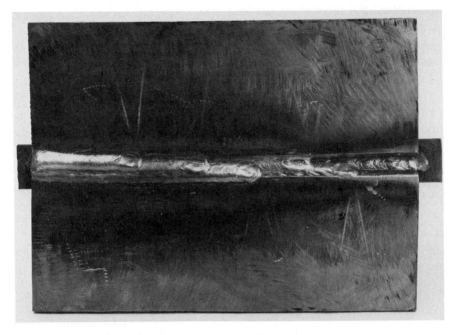

Figure 6-5 Low-hydrogen bead sequence

    Note: It is required that a short arc be held at all times when using the low-hydrogen electrode. Microporosity may result from a long arc.
2. Always deposit beads from the bottom of the joint to the top, taking advantage of the previous bead for a foundation to prevent sag or overlap.
3. Slag removal is very important. Remove slag carefully on every pass.

Procedure Without a Backing Strip E6010 and E6011

1. Clean all surfaces as suggested in Unit 3.
2. Set up and tack the plates together as shown in Figure 6-6.
3. Run the root pass with the electrode pointed upward, Figure 6-7. Use either the straight or stepping electrode motion as required to keep the key hole open and maintain complete root fusion.
4. Clean this bead and examine the root surface for incomplete fusion or excessive reinforcement.
5. Complete this weld in the same manner as the horizontal weld with a backing strip.

Procedure Without a Backing Strip E7018

1. Clean all surfaces as suggested in Unit 3.
2. Set up and tack the plates together as shown in Figure 6-6.

Single V-groove Horizontal Welds 2G on Plate

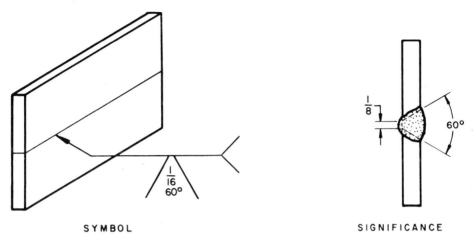

SYMBOL

SIGNIFICANCE

Figure 6-6 Horizontal welds on beveled plate without backing strip

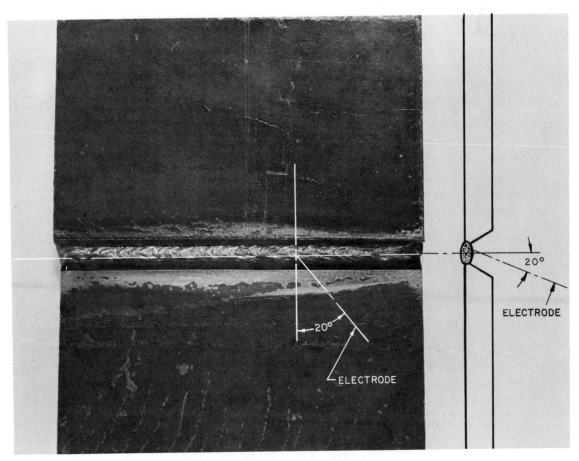

Figure 6-7 First pass on horizontal weld

Pipe Welding Techniques

3. Run the root pass with the electrode pointed upward, Figure 6-7. Use a straight electrode motion and a slightly higher than usual amperage setting to keep the key hole open and maintain complete root fusion.
4. Clean this bead and examine the root surface for incomplete fusion or excessive reinforcement.
5. Complete this weld in the same manner as the horizontal weld with a backing strip.

REVIEW

Select the letter preceding the best answer.

1. Why is uniform arc manipulation essential?
   a. To prevent the electrode from sticking
   b. To reduce the effect of arc blow
   c. To prevent spatter
   d. To produce uniform beads

2. How should the electrode be pointed for a horizontal weld?
   a. Slightly upward and into the weld pool
   b. Upward and ahead of the weld pool
   c. Downward and into the weld pool
   d. Straight into the weld pool

3. The most essential consideration in making the second pass for a horizontal weld with an E6010 electrode is to
   a. make a brief pause at the bottom of the weave.
   b. maintain a round molten puddle.
   c. clean the first bead.
   d. maintain a short arc.

4. Why is it important to produce uniform first and second beads?
   a. They can be produced in less time.
   b. They have a better appearance.
   c. They are less susceptible to undercut.
   d. They are easier to clean than irregular beads.

5. When making the final pass of the horizontal weld,
   a. make a three-count pause in the weave before reversing direction at the bottom of the weld.
   b. produce a wide weave bead.
   c. permit no irregularity in the appearance of the weld.
   d. strive for a circular shape of the crater.

Unit 7

SINGLE V-GROOVE OVERHEAD WELDS 4G ON PLATE

When pipe is welded in the horizontal fixed position, the bottom 160 degrees is considered to be welded in the *overhead* position, Figure 7-1. Since most cross-country pipe and a large proportion of all piping systems are welded in this position, the pipe welder can expect to be making nearly half of all welds in the overhead position. Practicing the overhead weld on plate is easier and less costly than practicing on pipe joints.

Note: A person unfamiliar with overhead welding should practice on a flat plate positioned 9 to 12 inches above eye level. If overhead welding is not new, proceed directly to the beveled butt joint in the following procedure.

Procedure with a Backing Strip E6010 and E6011

1. Set up the beveled plates as shown in Figure 7-2, and position them in a horizontal plane 9 to 12 inches above eye level or even higher if previous experience indicates a more desirable height.
2. Strike an arc and make the root pass. Be sure to make the bead wide enough to ensure good fusion at the root of both beveled plates.
3. Examine this bead for signs of slag inclusion and lack of fusion. Remove all slag before making a filler pass.

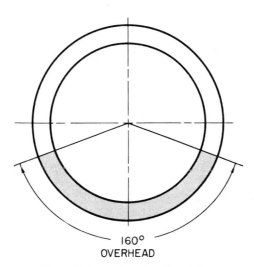

Figure 7-1 The overhead position

59

Pipe Welding Techniques

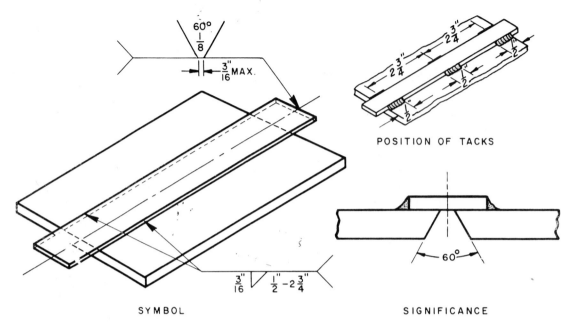

Figure 7-2 Overhead welds on beveled plate

Note: In overhead welding, slag tends to float upward into the root of the weld. This leaves gas and slag pockets that may not be visible until the specimens are machined or tested.

4. Make the second or filler pass. Manipulate the arc so that the finished bead has a relatively flat face.
   Note: Heavy, sagging beads cause slag to be trapped along the edges of the beads, Figure 7-4. This slag is difficult or impossible to remove. It becomes trapped by succeeding beads and causes the difficulties described in step 3.
5. Make the cap pass. Weave the arc so that the center of the electrode coincides with the edge of the bevel at each side of the weave. The finished weld need be no wider than the width of the top of the bevel plus the diameter of the electrode. Excessively wide beads waste time and material.
   Note: Undercutting is the result of either too long an arc or too short a pause at the sides of the weave. Slag inclusions and holes are caused by too long an arc, too fast a rate of advance, or low current settings.
6. Inspect the bead for undercut, holes, and slag inclusions. Use the precautions recommended to avoid these faults when making additional beveled butt joints in the overhead position.
7. Once this type of joint can be made free of all apparent defects, prepare two standard test specimens and test them in the guided-bend test jig.
   Note: When cutting these specimens, examine the cross section of each cut, looking for slag that may have been trapped in the weld. Occasionally, slag trapped at

Single V-groove Overhead Welds 4G on Plate

Figure 7-3 Position of hand and electrode holder for overhead welding

approximately the center of the weld does not show up in the guided-bend test and the weld passes. However, welders may be called upon to make X-ray quality welds. Slag pockets show up in X rays and disqualify the welder.

Procedure with a Backing Strip E7018

The techniques for using the low-hydrogen electrode in the overhead position are:
1. Stringer beads are used for refined grain structure and uniform consistency.
2. If a weave is used on successive passes after the first pass, it must be contained with a close pattern.
3. The electrode must be kept close to the work to produce a short arc and to insure proper shielding.

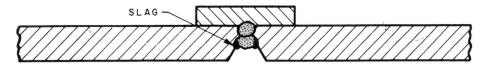

Figure 7-4 Slag trapped at edges of overhead bead with poor contour

Pipe Welding Techniques

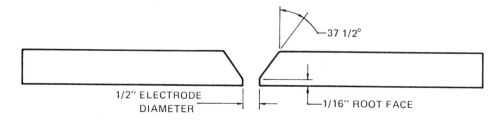

Figure 7-5 Plates beveled for overhead open root

Procedure Without a Backing Strip E6010 and E6011

1. Prepare the plates shown in Figure 7-5, and weld them in the overhead position. This joint corresponds to a pipe weld made without backing or chill rings.
   Note: Either place the tacks so that they are not in the area from which the test specimens are to be cut, or be extremely careful to clean all slag from both sides of the tacks before making the root pass.
2. Prepare test specimens from this joint as shown in Figure 7-6.
3. Place the specimen in a vise, break it sideways, and examine the fracture for signs of slag or gas pockets.

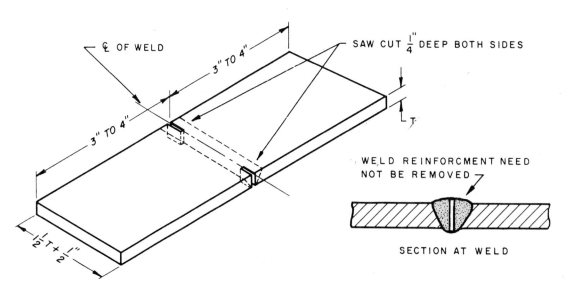

Figure 7-6 Specimen prepared for nick-break test

Single V-groove Overhead Welds 4G on Plate

REVIEW

Select the letter preceding the best answer.

1. The purpose of a nick-break specimen is to provide a test for
   a. tensile strength.
   b. ductility.
   c. porosity.
   d. elongation.

2. Why is cleaning between passes more important in the overhead position than other positions?
   a. Slag and gases float upward into the root of the weld instead of floating to the surface as in other positions.
   b. The slag runs downward.
   c. This position produces a wide molten pool.
   d. The molten pool can be more easily observed after cleaning in this position.

3. When making overhead welds, undercutting is
   a. less of a problem than when making horizontal welds.
   b. caused by too short an arc.
   c. the result of too long a pause at the sides of the weave.
   d. less of a problem than when making flat welds.

4. In overhead welding, the formation of icicles at the point at which the arc is broken can be prevented by
   a. using only a dual current machine.
   b. using a higher voltage.
   c. breaking the arc more rapidly.
   d. using a higher amperage.

Section 3

PIPE WELDING

Welded piping systems can be divided into three major categories. These categories are based on the pressure range or nature of the material that the system carries. The least critical systems are used to carry low pressure liquids such as water or for light structural applications such as hand rails. The medium-service systems are used to carry medium pressure liquids or gases such as residential natural gas or for such structural applications as highway sign posts. The heavy- or critical-piping systems are used for applications such as high-pressure steam, refineries and nuclear plants or for structural applications such as motorcycle and aircraft frames.

The welds in this section will stress the highest quality standards. It's important for the student to learn to be the best that there is in the field. Once the quality welds have been learned, the welder can make the required type of welds rapidly on the job when design allows it and production requires it.

Unit 8

BEADING

In this unit you will develop the skills required to make weld beads around pipe in all positions. The major difference is that in welding on pipe the work angle is constantly changing. As you make the transition from one position to another you must change the electrode angle. This continuous changing, in order to keep the work-to-electrode angle correct, is the unique part of pipe welding that makes it so difficult. Once you have mastered this skill then the rest of pipe welding should come easier.

MATERIALS

- One piece of 4-inch to 6-inch pipe
- 3/32 inch (2.3 mm) and 1/8 inch (3 mm) E6010, E6011 and E7018 electrodes
- Welding machine

Procedure Horizontal Rolled 1G E6010 and E6011

1. Place the pipe horizontally on the welding table.
2. Position yourself so that the pipe is directly in front of your body, Figure 8-1.
   Note: This position will give you good visibility of the weld and make it easier to follow a straight line.

Figure 8-1 Brace yourself so you are steady and comfortable.

65

Pipe Welding Techniques

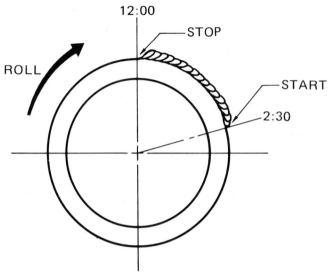

Figure 8-2 Pipe in the 1G horizontal rolled welding position

3. Strike the arc on the pipe in approximately the 2:30 position and weld to the 12:00 o'clock position, Figure 8-2.
   Note: Use both straight upward and stepping electrode motions.
4. Stop, chip, and inspect the weld for uniformity.
5. Roll the pipe so that the point where you stopped the arc is now at the 2:30 position. Strike the arc just ahead of the weld bead and, once the arc is stabilized, continue the weld as before.
6. Once the weld is completely around the pipe check to see that it is straight, uniform, and free of defects.
7. Repeat this welding procedure with both types of electrodes placing the weld beads so that they slightly overlap, Figure 8-3.

Procedure Horizontal Rolled 1G E7018

1. Place the pipe horizontally on the welding table.
2. Position yourself so that the pipe is directly in front of your body.
3. Strike the arc on the pipe in approximately the 2:30 position and weld to the 12:00 o'clock position.
   Note: Remember to strike the arc only on the pipe in a direct line with the weld to avoid unacceptable arc strikes that would not be covered up by the weld, Figure 8-4.
4. Stop, chip, and inspect the weld for uniformity.

Beading

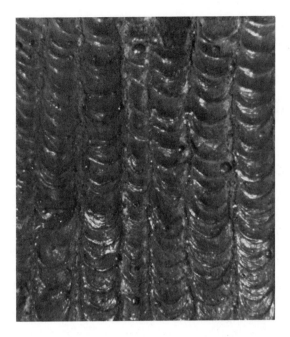

Figure 8-3 Practice stringer beads spaced properly on pipe.

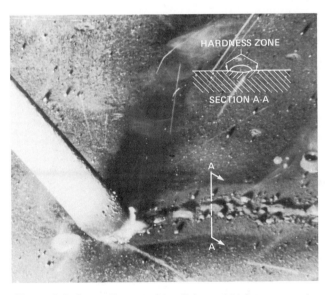

Figure 8-4 Arc strikes outside of the weld joint can result in hardness zones. (From Jeffus & Johnson, *Welding Principles and Applications.* Copyright 1984 by Delmar Publishers Inc.)

Pipe Welding Techniques

5. Roll the pipe so that the point where you stopped the arc is now at the 2:30 position. Strike the arc just ahead of the weld bead and, once the arc is stabilized, continue the weld as before.
6. Once the weld is completely around the pipe check to see that it is straight, uniform, and free of defects.
7. Repeat this welding procedure by placing the next weld bead so that they slightly overlap.

Procedure Horizontal Fixed Up 5G with E6010 and E6011

1. Place the pipe horizontally on the welding table at approximately chest level.
2. Position yourself so that you can comfortably move from the underside of the pipe to the top.
   Note: The pipe can be turned so that the back side can be more easily reached but it cannot be rolled, Figure 8-5.
3. Strike the arc on the pipe in approximately the 6:30 position and weld upward to the 12:00 o'clock position.
   Note: When it becomes necessary to stop because of an electrode change or you can no longer keep up with the angle change, try to taper the weld bead down in size, Figure 8-6.
4. Chip and inspect the weld for uniformity.
5. Strike the arc just ahead of the weld bead and, once the arc is stabilized, continue the weld as before.
6. Once the weld is completely around the pipe check to see that it is straight, uniform, and free of defects.
7. Repeat this welding procedure by placing the next weld bead so that they slightly overlap.

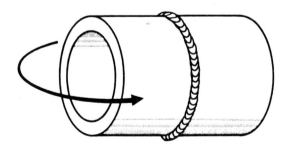

Figure 8-5 The pipe in the 5G position can be turned from front to back to make the backside easier to reach during training

Beading

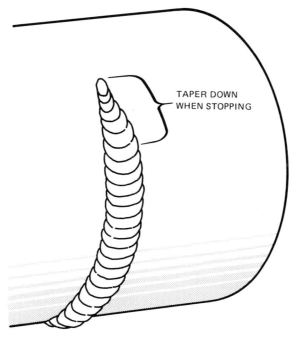

Figure 8-6 Taper the weld bead's size down before the weld is stopped

Procedure Horizontal Fixed Up 5G with E7018

1. Place the pipe horizontally on the welding table at approximately chest level.
2. Position yourself so that you can comfortably move from the underside of the pipe to the top.
3. Strike the arc on the pipe in approximately the 6:30 position and weld upward to the 12:00 o'clock position.
   Note: When it becomes necessary to stop because of an electrode change or you can no longer keep up with the angle change, try to taper the weld bead down in size, Figure 8-6.
4. Chip and inspect the weld for uniformity.
5. Strike the arc just ahead of the weld bead and, once the arc is stabilized, continue the weld as before.
6. Once the weld is completely around the pipe check to see that it is straight, uniform, and free of defects.
7. Repeat this welding procedure by placing the next weld bead so that they slightly overlap.

Pipe Welding Techniques

Procedure Horizontal Fixed Down 5G with E6010 and E6011

1. Place the pipe horizontally on the welding table at approximately chest level.
2. Position yourself so that you can comfortably move from the underside of the pipe to the top.
3. Strike the arc on the pipe in approximately the 12:00 o'clock position and weld downward to the 6:30 position, Figure 8-7.
4. Chip and inspect the weld for uniformity.
5. Strike the arc just ahead of the weld bead and, once the arc is stabilized, continue the weld as before.
6. Once the weld is completely around the pipe check to see that it is straight, uniform, and free of defects.
7. Repeat this welding procedure by placing the next weld bead so that they slightly overlap.

Procedure Vertical 2G with E6010 and E6011

1. Place the pipe vertically on the welding table at approximately chest level.
2. Position yourself so that you can comfortably move from one side of the pipe to the other, Figure 8-8.
3. Strike the arc on the pipe and make a weld around the pipe, Figure 8-9.
   Note: The pipe can be turned around to make it easier to weld on the backside but may not be turned from end to end.

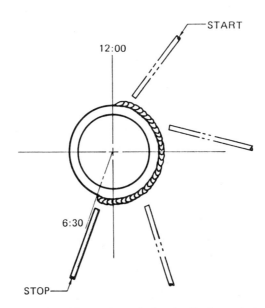

Figure 8-7 Electrode angle for the horizontal fixed 5G position

Beading

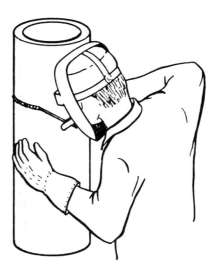

Figure 8-8 Position and brace yourself so you will be stable and have a clear view of the weld

4. Chip and inspect the weld for uniformity.
5. Strike the arc just ahead of the weld bead and, once the arc is stabilized, continue the weld as before.
6. Once the weld is completely around the pipe check to see that it is straight, uniform, and free of defects.
7. Repeat this welding procedure by placing the next weld bead so that they slightly overlap.

Procedure Vertical 2G with E7018

1. Place the pipe vertically on the welding table at approximately chest level.
2. Position yourself so that you can comfortably move from one side of the pipe to the other.
3. Strike the arc on the pipe and make a weld around the pipe.
   Note: E7018 electrodes have a weld that is somewhat more fluid and for that reason they may be difficult to use in this position.
4. Chip and inspect the weld for uniformity.
5. Strike the arc just ahead of the weld bead and, once the arc is stabilized, continue the weld as before.
6. Once the weld is completely around the pipe check to see that it is straight, uniform, and free of defects.
7. Repeat this welding procedure by placing the next weld bead so that they slightly overlap.

Pipe Welding Techniques

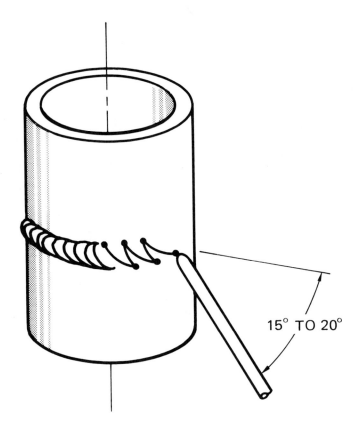

Figure 8-9 Electrode manipulations for vertical pipe 2G position

REVIEW

Select the letter preceding the best answer.

1. Pipe that is rolled horizontally during welding is in what position?
   a. 1G   b. 2G   c. 5G   d. 6G

2. Where should the arc be struck?
   a. Anywhere to light up the weld zone
   b. On a scrap plate to light up the weld zone
   c. Never outside the weld zone
   d. In a convenient location

3. When you have to stop the weld you should
   a. break the arc quickly.
   b. check to see if there are any hot sparks laying around.
   c. roll the pipe to the next position.
   d. taper the size of the weld bead down.

4. Pipe in the 5G position
   a. can be turned from the front to the back only.
   b. can be rolled from top to bottom.
   c. is like 3G plate and can only be turned as needed.
   d. can only be moved during welding.

5. Pipe welding is unique because
   a. only highly-skilled specialists can do it.
   b. the welding angle is constantly changing.
   c. the test for pipe are much stricter than for plate.
   d. All of the above.

Unit 9

FITTING THE BUTT JOINT

In this unit you will be learning some techniques for both preparing the end of the pipe and assembling the pipe joint for welding. One of the most critical elements of a good pipe weld is the preparation and assembly of the joint before any welding is actually started. In most pipe welding the welder is expected to do both the fitting and the welding. This allows welders to make the joint exactly the way they want to suit their individual skills and techniques within the code requirements. There are shops that have skilled craftsmen who just do the fitting; they are usually highly-skilled welders also.

END PREPARATION

There are a number of specialized devices used in the pipe-welding field for preparing the end of the pipe for welding. Some of these devices include air-powered pipe beveling tools, single- or double-torch beveling machines, and automatic cutting machines that are controlled by microprocessors, Figure 9-1, A, B, and C.

As a student you will probably use either a hand torch or a single-torch beveling machine. Some schools have shop-made devices such as the one in Figure 9-2 which uses an old car front spindle and some plate to make a turntable for cutting weld specimens.

Figure 9-1A Air-powered pipe beveling tool.
(Photo courtesy of E. W. Wachs Company)

Fitting the Butt Joint

Figure 9-1B Double torch holder on an H & M beveler. (Photo courtesy of H & M Pipe Beveling Machine Co., Inc.)

Figure 9-1C Computer-connected automatic cutting machine. (Photo courtesy of Vernon Tool Co.)

Pipe Welding Techniques

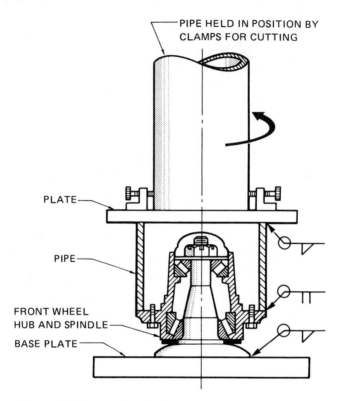

Figure 9-2 Turntable built from a front wheel assembly. (From Jeffus & Johnson, *Welding Principles and Applications.* Copyright 1984 by Delmar Publishers Inc.)

MATERIALS

- One or more pieces of 4-inch to 8-inch pipe
- Hand cutting torch
- Pipe cutting machine
- Hand grinder

Procedure Hand Beveling Pipe

1. Clamp the pipe securely in the horizontal position.
2. Mark a straight line around the pipe by following the edge of a flexible straightedge wrapped around the pipe.
3. Using a properly lit and adjusted hand cutting torch, make a square cut around the pipe following the line.
   Note: Be sure you are braced securely. You may cut the line by either holding the torch parallel or at a right angle to the centerline, Figure 9-3.

Fitting the Butt Joint

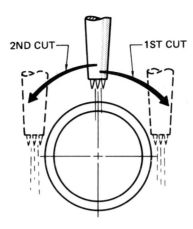

Figure 9-3 Pipe can be cut squarely if the torch angle does not change. (From Jeffus & Johnson, *Welding Principles and Applications.* Copyright 1984 by Delmar Publishers Inc.)

4. The squared end can now be beveled by either grinding or flame cutting, Figure 9-4.
    Note: If the bevel is cut using a hand torch its surface must be smoothed using a grinder, Figure 9-5.

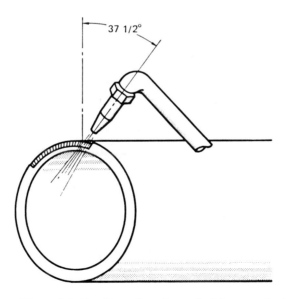

Figure 9-4 Hand beveling pipe end with an oxyfuel torch

Pipe Welding Techniques

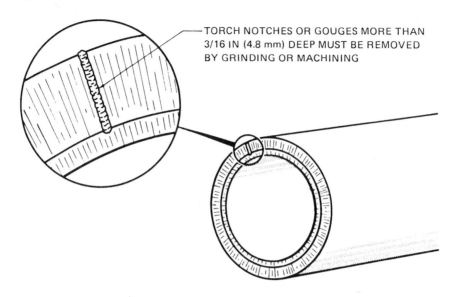

Figure 9-5 The pipe joint face must be smooth before welding

Procedure Machine Beveling Pipe

1. Clamp the pipe securely in the horizontal position.
2. Clamp the beveling machine securely to the pipe and crank the torch around to be sure it is at the correct height and that the hoses are free.
3. Light and adjust the torch. Allow the pipe to be preheated before slowly cranking the torch around.
   Note: Watch the cut to see that the kerf is smooth, Figure 9-6.
4. Remove the pipe cutting machine and grind the root face on the pipe end.

FITTING

Once the pipe end has been prepared it must be fitted to another pipe end to form the desired joint. If all pipe were perfectly round and the ends matched squarely then fitting would be easy. But all pipe is not round and squarely cut so some special skills are required to produce an acceptable joint. There are a number of specialty devices which can be used to align the pipe ends and other devices that can be used to check the alignment, Figure 9-7.

MATERIALS

- Two or more pieces of 4-inch to 8-inch beveled pipe
- Hand grinder

Fitting the Butt Joint

Figure 9-6 A correct cut. (From Jeffus & Johnson, *Welding Principles and Applications.* Copyright 1984 by Delmar Publishers Inc.)

- 3/32 inch (2.3 mm) or 1/8 inch (3 mm) E6010 or E6011 electrodes
- Welding machine

Procedure Fitting a Square Butt Joint

1. Place the ends of the pipes together and turn them to see if there is any position which provides the best fit.
2. Mark the pipe so that it can be separated to be worked on and then replaced in the same position.
3. Using a soapstone or other marker locate and mark the high spots that must be ground down to improve the fitting, Figure 9-8.
   Note: Work on only one pipe at a time and check to see that the alignments are improving.

Procedure Assembling the Butt Joint

1. Place the pipes in a V-block, angle iron, or other jig to hold the pipe in alignment so it can be welded, Figure 9-9.
2. Set the desired root spacing by using a gauge, welding wire, a coin, or other device, Figure 9-10.

Pipe Welding Techniques

Figure 9-7 Pipe alignment clamps. (Photo courtesy of G & W Products Inc.)

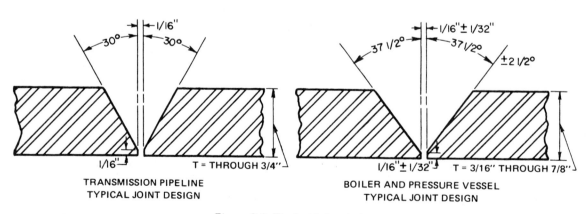

Figure 9-8 Typical joint design

Fitting the Butt Joint

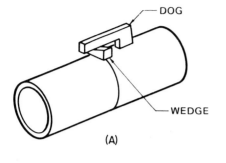
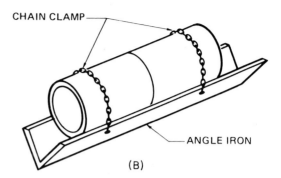

Figure 9-9 Three simple devices for aligning pipe. (A and B from Jeffus & Johnson, *Welding Principles and Applications.* Copyright 1984 by Delmar Publishers Inc.)

3. Using an electrode, carefully strike an arc in the joint and make a small tack weld, Figure 9-11.
   Note: If a pipe requires preheating before welding then it will also require preheating before tack welding.
4. Roll the pipe 180 degrees and make another tack weld.
5. Check to see that the pipe surfaces are aligned properly.
   Note: A hammer and anvil can be used to make slight adjustments in alignment if needed, Figure 9-12.

Pipe Welding Techniques

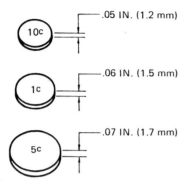

Figure 9-10 Root spacing provided by using a coin

Figure 9-11 Tack welding pipe joint

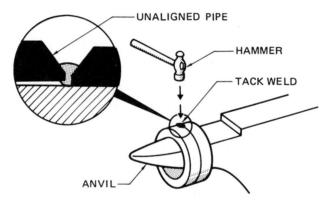

Figure 9-12 Hitting a hot tack weld can align a pipe joint. (From Jeffus & Johnson, *Welding Principles and Applications.* Copyright 1984 by Delmar Publishers Inc.)

6. Repeat the tacking procedure to produce tack welds at the desired locations, Figure 9-13.
7. Clean and inspect the tack welds for cracks or slag entrapments and repair if required.

REVIEW

Select the letter preceding the best answer.

1. Before the end of a pipe is torch beveled it must
    a. be ground smooth.
    b. be cut square.
    c. have all spatter removed.
    d. be made round.

Fitting the Butt Joint

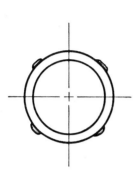
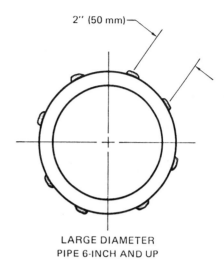

SMALL DIAMETER
PIPE 4-INCH AND UNDER

LARGE DIAMETER
PIPE 6-INCH AND UP

Figure 9-13 Avoid making tack welds at locations usually used for starting and stopping the weld beads

2. Pipe can be held in alignment for tack welding by a/an
   a. V-block.
   b. angle iron.
   c. alignment clamp.
   d. All of the above.

3. Grooves on the beveled surface larger than _____ must be ground out.
   a. 1/8 inch (3 mm)
   b. 3/16 inch (4.8 mm)
   c. 1/16 inch (.39 mm)
   d. 3/32 inch (2.3 mm)

4. A hammer and anvil are used for
   a. making the pipe round.
   b. smoothing the roughness of a tack out.
   c. major alignment of the out-of-roundness of the pipe.
   d. slight adjustments of alignment after tacking.

5. Tack welds on large-diameter pipe should be spaced
   a. approximately every 2 inches (50 mm).
   b. in six evenly-spaced places on the pipe.
   c. at the 12:00, 3:00, 6:00, and 9:00 o'clock positions.
   d. at the 2:00, 4:00, 8:00, and 10:00 o'clock positions.

Unit 10

WELDING HORIZONTAL PIPE 1G AND 5G POSITIONS

HORIZONTAL ROLLED 1G PIPE

The purpose of this unit is to introduce the student of pipe welding to the basic welded joint in its simplest form. The 1G position is the preferred welding position because welds in this position are faster and have fewer flaws than any other position.

This position is used in pipe fabrications when the section being welded is small enough to be rotated during welding. The positioning of the pipe can be done with either a manual or powered roller set or positioner, Figure 10-1.

MATERIALS

- Two or more pieces of 4-inch to 8-inch pipe
- 3/32 inch (2.3 mm) and 1/8 inch (3 mm) E6010, E6011 and E7018 electrodes
- Welding machine
- Hand grinder

Figure 10-1 Pipe rollers. (Photo courtesy of Atlas Welding Accessories, Inc.)

Welding Horizontal Pipe 1G and 5G Positions

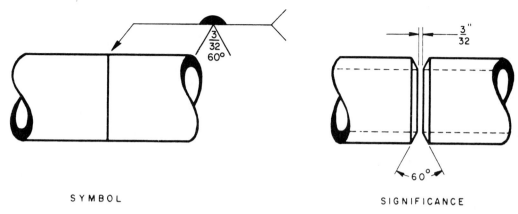

SYMBOL   SIGNIFICANCE

Figure 10-2 Horizontal pipe rolled

Procedure Open Root Pass with E6010 or E6011

1. Place a properly tack welded V-groove pipe joint on a table about chest high.
2. Start the weld at the 2:00 o'clock position and weld upward to the 11:00 o'clock position.
   Note: Use either the straight up or stepping motion to maintain the key hole just ahead of the weld bead, Figure 10-3.
3. Clean and inspect the weld before rotating the pipe to restart the weld. Continue this process until the weld is completely around the pipe.
4. After the root pass has been completed, clean and examine the bead for uniformity, slag holes, and excessive buildup at the center of the bead which might cause slag to be trapped along the edges of the weld. Inspect the inside of the pipe for excessive burn-through and the resulting icicles, as well as for lack of complete fusion at the root.

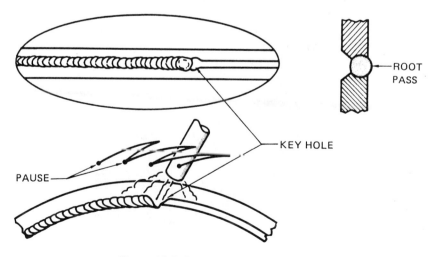

Figure 10-3 Stepping electrode motion

85

Pipe Welding Techniques

CAUSES OF DEFECTS IN THE PIPE WELD

- Slag inclusions: Slag left in the root opening when the electrode is moved too fast or too low a current is used.
- Excessive buildup: Poor timing of the arc cycle; too low or high a current used.
- Excessive burn-through: Excessive current or too large a root opening.
- Undercut: Excessive travel speed; incorrect electrode angle; too long an arc or too high a current setting.

Procedure Root Pass Cleanup

1. Using a pipe with a root pass, clamp it in a vise at a comfortable height.
2. Locate any welding defects after the joint has been cleaned of slag by using a chipping hammer and a wire brush.
3. Using a hand grinder, as shown in Figure 10-4, lightly grind out any defects and any slag trapped along the toe of the root pass.
   Note: Be sure to use a pipe grinding disk because it is thinner and will not widen the joint, Figure 10-5.

Figure 10-4 Hold the grinding stone so it does not widen the groove

Welding Horizontal Pipe 1G and 5G Positions

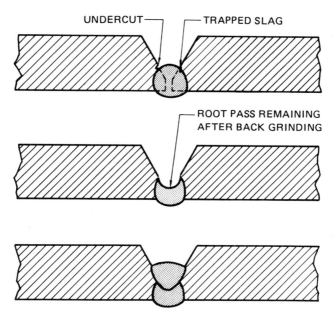

Figure 10-5 Back grinding to remove discontinuities. (From Jeffus & Johnson, *Welding Principles and Applications.* Copyright 1984 by Delmar Publishers Inc.)

Procedure Open Root Pass with E7018

1. Place a properly tack welded V-grooved pipe joint on a table about chest high.
2. Start the weld at the 2:00 o'clock position and weld upward to the 11:00 o'clock position, Figure 10-6.
3. Clean and inspect the weld before rotating the pipe to restart the weld. Continue this process until the weld is completely around the pipe.
4. Clean and inspect the weld before continuing with the final passes.

Procedure Filler and Cover Pass 1G

1. Proceed with the second bead, starting at the 2:00 o'clock position and using the weave motion. Rotate the pipe and clean the crater at each change of position. Again, remember to modify the weave motion as the weld approaches the flat position.

    Note: This second bead should complete the weld on 4-inch standard pipe. It should be inspected and all faults should be noted and corrected when additional pipe joints of this type are made. Since this is the final or cap pass, special attention should be given to any evidence of undercutting. Undercutting is caused by too short a pause at the sides of the weave or too long an arc in the same area. Un-

Pipe Welding Techniques

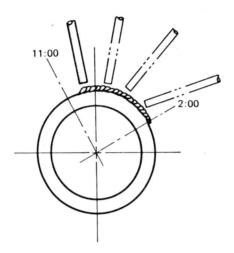

Figure 10-6 E7018 electrodes do not use as great an upward angle

dercutting causes a concentration of stresses in the undercut area and frequently leads to a rupture starting at this point. The same hazard exists when fusion is not complete at the root of the joint.
2. Obtain more beveled pipe and continue to make rolling pipe welds until the welded joints are of consistently good quality in both the root pass and cap pass. Welds of this type may be submitted to X-ray inspection which detects even the most minute defects. A good rolling pipe weld should have the contour and appearance shown in Figure 10-7.
3. When acceptable welds can be produced in every joint, cut two test specimens and test them in the guided-bend test jig.
4. Examine the root-bend specimen carefully for any signs of lack of fusion or slag holes. Any fault more than 1/8 inch long in any direction constitutes a failure.
5. Examine the face-bend specimen for signs of undercutting at the line of fusion and for any signs of slag pockets or holes in the face of the weld.
6. If failure is indicated after testing the specimens, make more welds in this position, correcting for the faults until each weld can pass the guided-bend jig test.

HORIZONTAL FIXED 5G PIPE

In this section you will learn both the vertical-up and the vertical-down welding techniques, even though pipe welding and pressure vessel welding do not use the vertical-down method as widely as vertical-up. For that reason, you should spend more time practicing the vertical-up weld.

Welding Horizontal Pipe 1G and 5G Positions

Figure 10-7 Acceptable rolling pipe joint

Procedure Open Root Pass with E6010 or E6011 Vertical-up 5G

1. Place a properly tack welded V-grooved pipe joint on a table about chest high, Figure 10-8.
2. Start the weld at the 6:30 position and weld upward toward the 11:30 position, Figure 10-9.
   Note: Use either the straight up or stepping motion to maintain the key hole just ahead of the weld bead.
3. Clean and inspect the weld before restarting the weld on the other side.
4. Locate any welding defects after the joint has been cleaned of slag by using a chipping hammer and a wire brush.

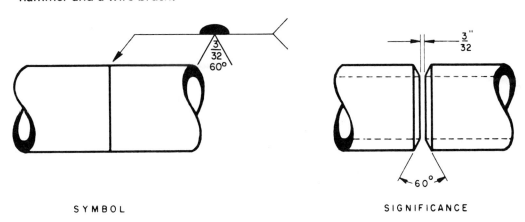

SYMBOL   SIGNIFICANCE

Figure 10-8 Fixed-position butt welds on horizontal pipe

Pipe Welding Techniques

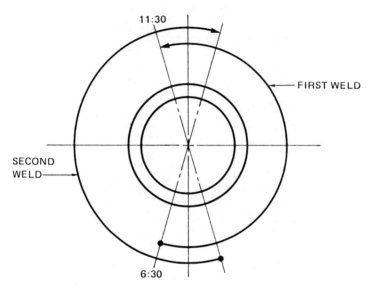

Figure 10-9 Starting and stopping points for the first and second welding passes

5. Using a hand grinder lightly grind out any defects and any slag trapped along the toe of the root pass.
6. Continue to make welds in the groove until it is filled, Figure 10-10.

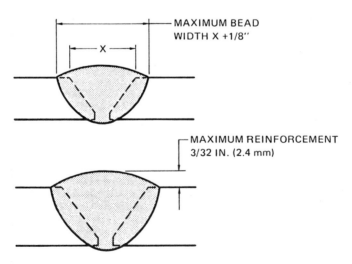

Figure 10-10 Maximum bead width and reinforcement for 3/8 (10 mm) wall thickness and less pipe

Welding Horizontal Pipe 1G and 5G Positions

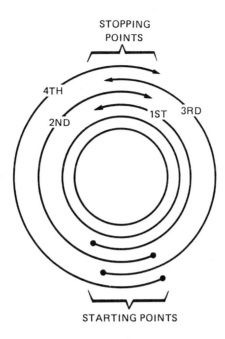

Figure 10-11 For multiple-pass welds stagger the starting and stopping points

7. Continue to make this type of joint until each weld is of good appearance. It is not necessary to bend-test every weld made.
   Note: Stagger the locations of the stopping and starting points, Figure 10-11. The reason for starting and stopping beads at about 30 degrees off center is illustrated in Figure 10-12. This drawing shows the order of removal of test specimens from pipe welds. The probability of slag pockets, holes, or lack of complete fusion is always greatest where a weld is started or completed. If these points are kept outside of the area of the test specimens, the chances of failure are decreased.

Procedure Open Root Pass with E7018 Vertical-up 5G

1. Place a properly tack welded V grooved pipe joint on a table about chest high.
2. Start the weld at the 6:30 position and weld upward toward the 11:30 position.
   Note: Use the straight up motion with a slight side-to-side rocking of the electrode holder to maintain the key hole just ahead of the electrode, Figure 10-13.
3. Clean and inspect the weld before restarting the weld on the other side.
4. Locate any welding defects after the joint has been cleaned of slag by using a chipping hammer and a wire brush.

Pipe Welding Techniques

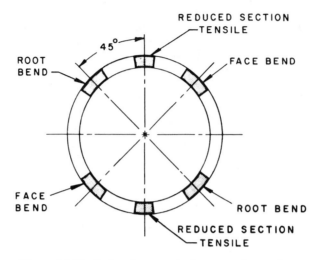

Figure 10-12 Order of removal of test specimens for pipe 1/16"–3/4" thick—as shown. For pipe over 3/4" cut side bend specimens from indicated 45-degree areas.

5. Using a hand grinder lightly grind out any defects and any slag trapped along the toe of the root pass.
6. Continue to make welds in the groove until it is filled and test it if it appears to be good.

Procedure Open Root Pass with E6010 or E6011 Vertical-down 5G

1. Place a properly tack welded V-grooved pipe joint on a table about chest high.
2. Start the weld at the 11:30 position and weld downward toward the 6:30 position, Figure 10-14.

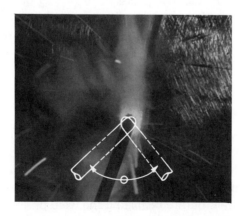

Figure 10-13 Rocking the electrode will help the weld to fuse along the side

Welding Horizontal Pipe 1G and 5G Positions

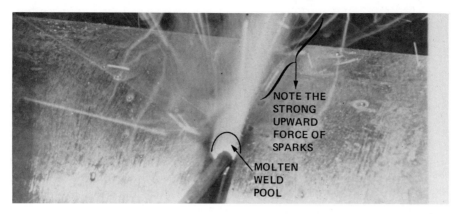

Figure 10-14 Watch the molten weld pool shape in the vertical-down position

Note: Use a very short arc drag-type motion. If the key hole becomes too large use a short back step to close it up, Figure 10-15. If this fails to close the root opening then it will be necessary to weld this joint in the vertical-up position.
3. Clean and inspect the weld before restarting the weld on the other side.
4. Locate any welding defects after the joint has been cleaned of slag by using a chipping hammer and a wire brush.
5. Using a hand grinder lightly grind out any defects and any slag trapped along the toe of the root pass.
6. Continue to make welds in the groove until it is filled and test the weld if it appears to be acceptable.

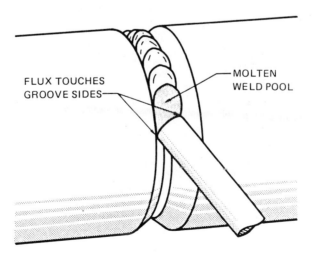

Figure 10-15 Electrode position for a vertical-down weld

93

Pipe Welding Techniques

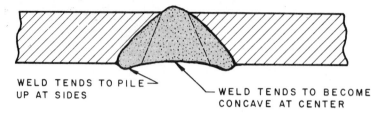

Figure 10-16 Cross section of vertical-down weld

Note: Each joint should be examined both inside and outside for lack of fusion, holes, slag inclusions, undercutting, and the contour of the weld. Inspection of the contour is important when vertical-down welding this type of joint, because the bead tends to be concave, Figure 10-16. Concave beads make the weld thinner at the center than the wall thickness of the pipe and, naturally, present an area weaker than the surrounding material. This fault can be corrected by using a slightly slower weave motion or by using slightly less amperage at the arc.

Note: It is important that this electrode angle remain constant while going around the

Figure 10-17 Electrode angle

Welding Horizontal Pipe 1G and 5G Positions

pipe. Variations in the proper angle may cause excessive burn-through, holes, and loss of arc. This type of joint should be set up with a slightly wider than normal root opening to compensate for the shorter arc length if the bevel has a land such as shown in Figure 10-17.

REVIEW

Select the letter preceding the best answer.

1. Why not proceed beyond eleven o'clock with the weld—for example, to ten o'clock or nine o'clock?
   a. Welds beyond eleven o'clock would be considered vertical-down and would require higher amperage.
   b. Cleaning the crater would be more difficult.
   c. The tacks would interfere with the weld.
   d. Excessive burn-through would be produced.

2. In the root pass, poor root fusion is caused by
   a. poor timing of the arc cycle.
   b. too high an amperage.
   c. too close a root spacing.
   d. too great a root spacing.

3. Which type of welded joint is encountered most frequently in pipeline work?
   a. Horizontal fixed position
   b. Rolling pipe joint
   c. Vertical fixed position
   d. 50 percent horizontal, 50 percent vertical

4. If the root opening is found to be too large when welding vertical-down you should
   a. lower the amperage.
   b. lower the electrode angle.
   c. weld faster.
   d. weld vertical-up.

5. The maximum weld reinforcement for a 3/8 inch (10 mm) wall thickness pipe should be
   a. 3/32 inch (2.4 mm).
   b. 1/8 inch (3 mm).
   c. 1/16 inch (.39 mm).
   d. the higher the better.

6. How can the key hole be maintained just ahead of the weld when welding vertical up?
   a. Use a stepping motion.
   b. Change electrode size.
   c. Keep the electrode angle at 90°.
   d. Change the type of electrode used.

Pipe Welding Techniques

7. To avoid excessive weld reinforcement 4-inch standard pipe is usually welded using how many passes?
   a. 1 b. 2 c. 3 d. 4

8. The probability of having slag pockets or holes is greatest when
   a. at the bottom of the pipe.
   b. using AC welding current.
   c. when starting a new electrode.
   d. using DCSP welding current.

9. How does a vertical-down pipe weld tend to be shaped?
   a. It piles up along the side and is concave in the center.
   b. It is convex in the center and underfills the sides.
   c. It undercuts the toe and overfills the center.
   d. It has a smooth transition to an ideal convex center.

10. How can undercutting be controlled on vertical-up welds?
    a. Use a longer arc length.
    b. Use a faster travel rate.
    c. Try to stay in the center of the weld.
    d. Pause longer at the sides of the weld.

Unit 11

WELDING VERTICAL PIPE 2G POSITION

The student of pipe welding should gain proficiency in making fixed-position butt welds on vertical pipe. This is probably the second most frequently used joint when welding pipe in the field. This position has been labeled the 2G test position by the ASME code.

MATERIALS

- Two or more pieces of 4-inch to 8-inch pipe
- 3/32 inch (2.3 mm) and 1/8 inch (3 mm) E6010, E6011 and E7018 electrodes
- Welding machine
- Grinder

Procedure Open Root with E6010 or E6011

1. Align the pipe and tack it in place in a vertical position at eye level.
2. Most of the steps in making horizontal joints in vertical pipe are similar to those for horizontal welds on vertical plate.
3. Adjust the machine to normal current. Make the first pass with the electrode pointing toward the longitudinal centerline of the pipe at all times and pointed upward at a 15- to 20-degree angle, Figure 11-2.

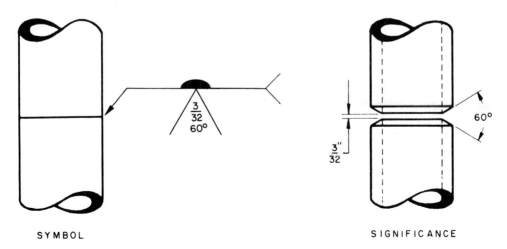

SYMBOL   SIGNIFICANCE

Figure 11-1 Horizontal butt welds on vertical pipe

97

Pipe Welding Techniques

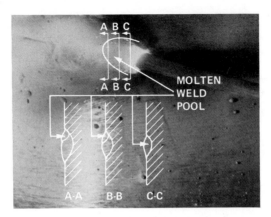

Figure 11-2 The molten weld pool rests on the solid weld metal below

    a. Examine the crater and the finished bead for signs of excessive burn-through, sag, and undercut as the weld progresses. Correct these faults by using a shorter arc and/or a higher rate of travel.

    b. As the weld progresses the operator's position must change in order to maintain the proper rod angle in relation to the centerline of the pipe.

    c. When pausing to change position, clean the bead and crater thoroughly. Inspect the inside of the joint for lack of complete fusion and excessive burn-through.

    d. Correct for incomplete fusion by using a slightly higher current setting or a slightly lower rate of travel.

    e. Correct for excessive burn-through by using less current, a higher rate of travel, a shorter arc, or a combination of these.

    f. When continuing this bead, start the arc one inch ahead of the crater and return to the crater with a long arc. Reduce the arc to the proper length. This ensures good fusion in the crater and eliminates the possibility of excessive buildup at the starting point.

4. After the root pass has been completed, clean and inspect the bead both inside and outside. Observe all faults and correct them when making additional root passes.
5. Make the filler pass. Compensate for the curvature of the pipe so that the electrode is always pointing toward the centerline of the pipe and upward at a 15- to 20-degree angle.
6. When pausing to change position, clean and inspect the bead. Be critical of any undercut or sag. Undercutting may cause slag pockets when making the cap pass, and sag may cause poor fusion or appearance.

    a. Compensate for undercutting by a longer pause at the top of the weave or by using a shorter arc in the same area.

    b. Eliminate sag by using a greater angle for the weave, a more rapid advance toward the bottom of the weave, or a reduction in amperage at the arc. Figure 11-3 indicates the weave motion, angle, and pause.

Welding Vertical Pipe 2G Position

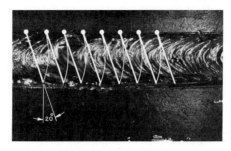

Figure 11-3 Weave motion for filler pass

7. After the filler pass has been completed, cleaned, and inspected, make the final pass as a weave bead.
   Note: Start the final pass at some point other than the starting point of the filler pass. Any tendency toward buildup at the start will naturally be increased if the subsequent beads are started in the same area. Figure 11-4 shows a completed horizontal pipe weld on vertical pipe with the weave pattern and pause indicated.
8. Clean and inspect the completed weld. If visual inspection indicates a satisfactory joint, cut, prepare, and test specimens removed from the four sides, Figure 11-5.

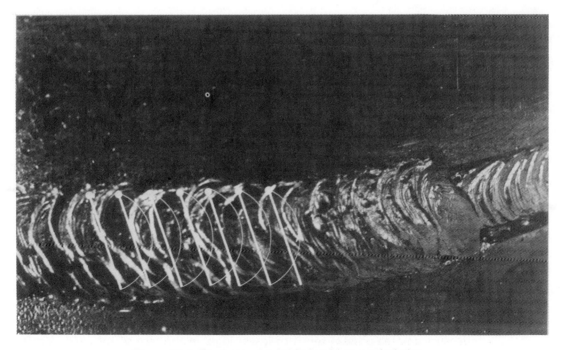

Figure 11-4 Final pass—horizontal weld on vertical pipe

Pipe Welding Techniques

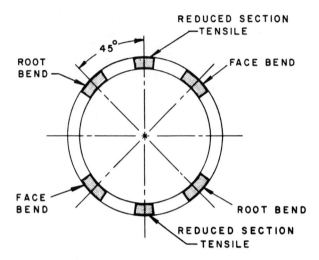

Figure 11-5 Order of removal of test specimens for pipe 1/16"-3/4" thick—as shown. For pipe over 3/4" cut side bend specimens from indicated 45-degree areas.

Note: If faults such as undercutting, sag, incomplete fusion, or holes in the face of the weld are noted, continue to make welds of this type until they pass visual inspection before making additional bend tests.

Procedure Open Root with E6010 or E6011 Electrodes Stringer Beads

1. Place the properly tack welded V-grooved pipe joint in the vertical position at a comfortable height.
2. Make the root pass as before and check for soundness.
3. Make the second, third, fourth, fifth, and sixth weld beads in order as indicated in Figure 11-6.
4. Be sure to clean each bead before applying the succeeding bead. Keep the electrode pointed at the centerline of the pipe at all times and slanted upward 10 to 15 degrees. The precaution about starting each bead in a slightly different area is more important for this type of joint than others because of the total number of beads.
5. Cut specimens from the finished joint and follow the standard test procedure.
6. Cut additional specimens, polish, and macroetch them, following the procedure for macroetching in Unit 3. Compare the grain structure of the etched specimen with one cut from a pipe weld made with a weave pattern.
7. Make additional welds with stringer beads, but apply beads 2 and 4 at the bottom of the V instead of the top.
    Note: This may lead to a somewhat rougher bead than those produced in steps 3 and 5 because any sag present tends to leave ridges in the face of the weld. A smaller

Welding Vertical Pipe 2G Position

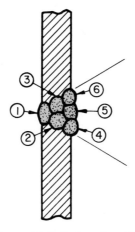

Figure 11-6 Order of making beads for horizontal butt welds

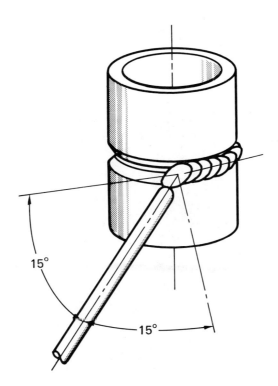

Figure 11-7 Electrode angle for horizontal welds

Pipe Welding Techniques

electrode angle (5 to 10 degrees), a slightly higher rate of travel, and a slightly higher current setting helps to overcome the sag and resulting roughness.
8. Prepare test specimens from this joint and test in the same manner.

Procedure Open Root with E7018 Electrodes

1. Place the properly tack welded V-grooved pipe joint in the vertical position at a comfortable height.
2. Make the root pass using both a slightly upward (15 degree) and slightly trailing (15 degree) angle, Figure 11-7.
   Note: If the root opening is very small then a slightly higher welding current can be used with a very short arc length to increase the quality of the root pass.
3. Take time to clean the weld crater each time the electrode is changed or the arc is stopped.
4. Cleaning the root pass when an E7018 electrode is used often requires more work because the slag sticks tightly to the weld.
   Note: A hand grinder with a wire wheel will help to clean the slag, Figure 11-8.
5. Lower the current and continue by making the filler and cover passes as small stringer beads, Figure 11-9.
6. Test the weld to see if it will pass the guided-bend test and repeat this procedure if necessary.

Figure 11-8 Wire brushes and grinding stones used to clean up welds

Welding Vertical Pipe 2G Position

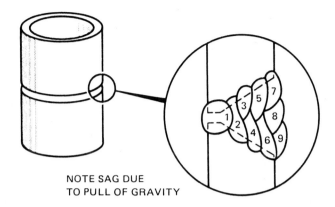

Figure 11-9 Multipass weld bead positions for a horizontal weld (From Jeffus & Johnson, *Welding Principles and Applications.* Copyright 1984 by Delmar Publishers Inc.)

REVIEW

Select the letter preceding the best answer.

1. When possible, horizontal welds on material in a vertical position should be
   a. waist high.
   b. flat on the table.
   c. inclined at 45 degrees.
   d. at eye level.

2. Correct for faults as the weld progresses by using a
   a. shorter arc if there is excessive burn-through.
   b. lower rate of travel if there is undercutting.
   c. longer arc is there is sag.
   d. longer arc if there is excessive burn-through.

3. In making the root pass, correct for incomplete fusion by
   a. changing the electrode angle to 30 degrees.
   b. using a slightly higher rate of travel.
   c. starting the arc in the crater.
   d. using a slightly higher current.

4. In making the filler pass, correct faults by
   a. compensating for the curvature of the pipe by pointing the electrode toward the centerline of the pipe.
   b. reducing undercutting by a shorter pass at the top of the weave.
   c. reducing undercutting by a longer arc at the top of the weave.
   d. eliminating sag by an increase in amperage.

Pipe Welding Techniques

5. When making the cap pass,
    a. begin at the starting point of the filler pass.
    b. take care in regard to the appearance of the bead.
    c. keep the center of the electrode even with the top of the weave.
    d. begin at the starting point of the root pass.

Unit 12

WELDING 45-DEGREE INCLINED PIPE 6G AND 6GR POSITIONS

Passing the 6G test will meet the code requirements for qualification or certification of welders. The weld around a pipe in this position incorporates part of all the welding positions, Figure 12-1. To pass the welding test for this position the welder must be highly skilled in the other positions.

The most difficult of all pipe welds is the 6GR. The pipe is fixed in the same 45 degree incline as the 6G but there is a restricting ring placed just above the weld, Figure 12-2. This ring reduces accessibility to the weld joint. The ability to pass this test demonstrates the highest degree of welding skill.

MATERIALS

- Two or more pieces of 4-inch to 8-inch pipe
- 3/32 inch (2.3 mm) and 1/8 inch (3 mm) E6010, E6011 and E7018 electrodes
- Welding machine
- Hand grinder

Procedure Open Root Pass with E6010 or E6011 and E7018 6G

1. Place a properly tack welded V-grooved pipe joint on a table at a 45-degree fixed angle.
2. Start the root pass near the bottom of the joint and weld upward.
   Note: It may be necessary to adjust the current setting as the weld progresses around the pipe.

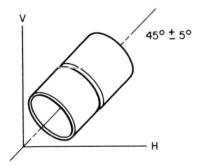

Figure 12-1 Test position 6G

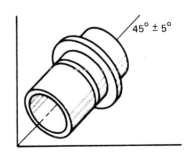

Figure 12-2 Test position 6GR

105

Pipe Welding Techniques

3. After the root pass is completed clean and grind it as necessary.
4. Using a stringer bead technique continue with the filler and cover passes until the joint is completed.
   Note: Be sure to clean each weld pass thoroughly to prevent slag entrapments.
5. Cut specimens from the finished joint and follow the standard test procedures.
6. Repeat this procedure using the E7018 electrode.

Procedure Open Root Pass with E6010 or E6011 and E7018 6GR

1. Place a properly tack welded V-grooved pipe joint on a table at a 45-degree fixed angle, Figure 12-3.
2. Start the root pass near the bottom of the joint and weld upward.
   Note: The restricting ring may require that you get your welding helmet in closer so you can see.
3. After the root pass is completed clean and grind it as necessary.
4. Using a stringer bead technique continue with the filler and cover passes until the joint is completed.
5. Cut specimens from the finished joint and follow the standard test procedures.
6. Repeat this procedure using the E7018 electrode.

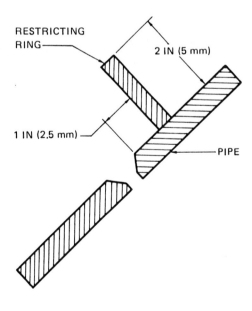

Figure 12-3 Location of the restricting ring

Welding 45-degree Inclined Pipe 6G and 6GR Positions

REVIEW

Select the letter preceding the best answer.

1. The 6G welding position is a combination of
   a. the flat and horizontal welding positions.
   b. the horizontal and vertical welding positions.
   c. the vertical and overhead welding positions.
   d. all welding positions.

2. The inclined angle of most joints for this positions is
   a. 60 degrees.
   b. 75 degrees.
   c. 45 degrees.
   d. 100 degrees.

3. The electrode used when welding pipe in this position is
   a. E6010.
   b. E6018.
   c. E7018.
   d. not specified.

4. Qualifying on pipe in the 6G position will meet code requirements for which of the other position(s)?
   a. 1G
   b. 2G
   c. 5G
   d. All of the above.

5. The 6GR root pass should be started
   a. near the bottom.
   b. at the top.
   c. on the side.
   d. It's up to the welder to choose the starting point.

Section 4

SPECIAL APPLICATIONS

To be an excellent pipe welder you must be able to do other things beside SMA weld pipe joints. The highly-skilled craftsman will be able to use the GTA, GMA and FCA welding processes to produce welded joints and fabricate some special fittings.

These specialty skills are well beyond the responsibilities of the new welder. These jobs are usually held aside for the more experienced welders as a reward for their efforts. You will find that it often takes time for the "new welder" to receive the trust to be allowed to do these jobs. In this section you will be learning the essentials to be able to prove yourself on the job when the opportunity arrives.

Unit 13

GAS TUNGSTEN ARC WELDING OF PIPE

The gas tungsten arc welding (GTAW) process is used for pipe welding because of a number of its qualities. The GTAW process can be used on a very wide range of metals or their alloys, thicknesses, and positions. This extreme flexibility and the high quality of the welds produced has resulted in this process being selected when either manual or automatic high-integrity welding is required, Figure 13-1.

GTA welding may be used only to make the root pass with or without a backing ring or it may be used to produce the complete weld. In the procedures in this unit you should evaluate the root pass as a separate weld because it is so often separate in industry.

MATERIALS

- One or more pieces of 4-inch to 8-inch pipe
- A complete GTAW setup

Figure 13-1 Automatic pipe welding unit for large diameter pipe. (Photo courtesy of Hobart Brothers Co.)

Pipe Welding Techniques

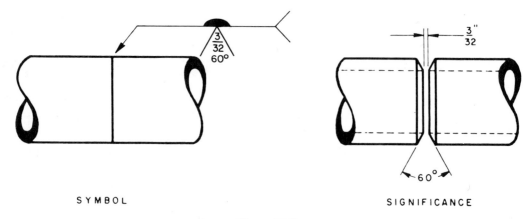

SYMBOL

SIGNIFICANCE

Figure 13-2

- 1/16 inch (1.5 mm), 3/32 inch (2.3 mm), and 1/8 inch (3 mm) RG60 or E70S-3 cut length welding rods

Procedure Open Root Pass 1G Position

1. Place a properly tack welded V-grooved pipe joint on a table about chest high, Figure 13-2.
2. Set the tungsten extension so that when the gas cup is resting on the sides of the V-groove the tungsten can be held near the root, Figure 13-3.
3. You will start the weld at the 2:30 position and weld upward. Put the cup against the V-groove and lower the filler rod into the weld zone.

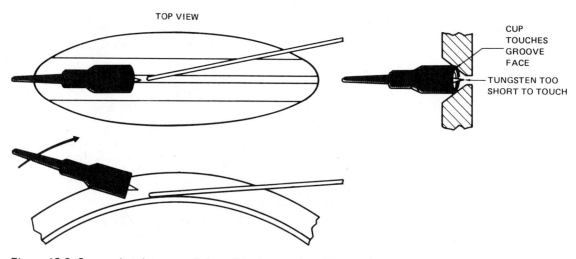

Figure 13-3 Supporting the cup and the rod in the groove will help when making the root pass. (From Jeffus & Johnson, *Welding Principles and Applications.* Copyright 1984 by Delmar Publishers Inc.)

Gas Tungsten Arc Welding of Pipe

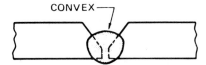

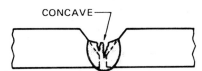

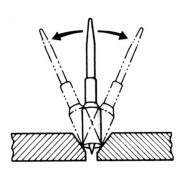

Figure 13-4

Figure 13-5 Inadequate root pass size can cause centerline cracking of the weld bead.

4. Lower your welding hood and depress the foot control to start an arc.
5. Watch the sides of the root opening to see when they begin to melt. You may have to rock the torch to both sides to get uniform melting, Figure 13-4.
    Note: Do not add filler until both sides of the root have melted.
6. Once you have a molten pool at the root add the filler to the top leading edge of the pool as the tungsten is withdrawn slightly.
    Note: You must add enough filler metal to make a weld that is slightly convex, Figure 13-5.
    Note: The tungsten can be moved back by slightly rocking the torch to one side, Figure 13-6.

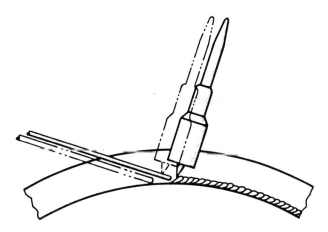

Figure 13-6 The torch moves back as the filler rod is added to the weld

Pipe Welding Techniques

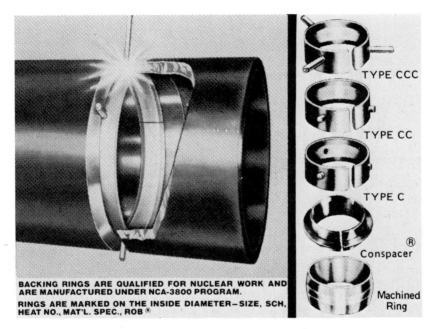

Figure 13-7 Specialized backing rings. (Reproduced by permission of Robvon Backing Ring Co.)

7. When it becomes necessary to stop, either taper the weld pool size down or leave the filler rod stuck in the weld crater. Either of these techniques will reduce the possibility of crater cracking.
8. Continue the root pass around the pipe. Clean and inspect it for flaws.

PIPE WELDING WITH INTERNAL RINGS

Some pipe welding procedures require the use of various rings within the pipe. The different designs were created to serve varying functions.

The *backing* or *chill ring,* Figure 13-7, is used when specified to back up the joint and help eliminate burn-through on the root pass. This type of ring is also used as a chill ring or heat sink, to help prevent burn-through, especially on lightweight tubing. The disadvantage of using this type of ring is the restriction it creates within the pipeline. It is preferred that the pipeline have a smooth inner surface.

To eliminate the restriction created by the standard rings, many specifications now require the use of *consumable insert rings,* Figure 13-8. This ring is designed to be melted into the inside edges of the pipe with the GTAW process, using only the tungsten arc as the heat source. After running the root bead, the inner surface of the weld should present a slight convex surface with no undercut. During the welding operation, the inside of the pipe is

Gas Tungsten Arc Welding of Pipe

Figure 13-8A Consumable insert rings. (Reproduced by permission of Robvon Backing Ring Co.)

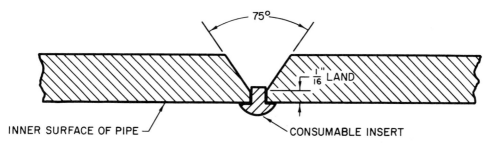

Figure 13-8B A consumable insert ring in position

sometimes purged with the inert gas, argon. This is done by sealing off the pipe and forcing argon into the pipe pushing out atmospheric gases. The argon will then protect the root side of the bead from oxidation, Figure 13-9.

Procedure Root Pass 1G Position with a Consumable Insert

1. Place a properly tack welded V-grooved pipe joint on a table about chest high, Figure 13-10.
2. Set the tungsten extension so that when the gas cup is resting on the sides of the V-groove the tungsten can be held near the root.
3. You will start the weld at the 2:30 position and weld upward. Put the cup against the V-groove.

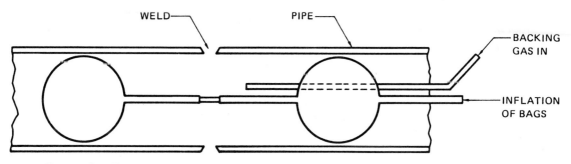

Figure 13-9 Two inflatable bags can be used to seal a pipe so that a backing gas can be used. (Reproduced by permission of Safety Main Stopper Co., Inc.)

Pipe Welding Techniques

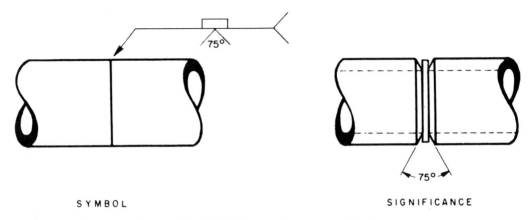

SYMBOL SIGNIFICANCE

Figure 13-10 Weld joint with a consumable insert

4. Lower your welding hood and depress the foot control to start an arc.
5. Watch the sides of the root opening to see when they begin to melt. You may have to rock the torch to both sides to get uniform melting of both sides of the root and the backing ring.
6. Once you have a molten pool at the root use a rocking motion of the cup against the sides of the V-groove to walk the torch up the joint.
7. When it becomes necessary to stop, taper the size of the weld pool down.
8. Continue the root pass around the pipe. Clean and inspect it for flaws.

Procedure Open Root Pass 5G Position

1. Place a properly tack welded V-grooved pipe joint on a table about chest high.
2. Set the tungsten extension as before.
3. You will start the weld at the 6:30 position and weld upward. Put the cup against the V-groove and lower the filler rod into the weld zone.
4. Lower your welding hood and depress the foot control to start an arc.
5. Watch the sides of the root opening to see when they begin to melt. You may have to rock the torch to both sides to get uniform melting.
6. Once you have a molten pool at the root add the filler to the top leading edge of the pool as the tungsten is withdrawn slightly.
   Note: You must add the filler by touching it into the molten weld pool. The surface tension of the weld pool will hold it in place so the tungsten does not become contaminated.
7. When it becomes necessary to stop, taper the size of the weld pool down or leave the filler rod stuck in the weld crater.
8. Continue the root pass around the pipe. Clean and inspect it for flaws.

Gas Tungsten Arc Welding of Pipe

Procedure Open Root Pass 2G Position

1. Place a properly tack welded V-groove pipe joint on a table about chest high.
2. Set the tungsten extension as before.
3. You will start the weld nearest to yourself and weld around the pipe away from yourself, Figure 13-11.
4. Lower your welding hood and depress the foot control to start an arc.
5. Watch the sides of the root opening to see when they begin to melt. You may have to rock the torch to both sides to get uniform melting.
6. Once you have a molten pool at the root add the filler to the top leading edge of the pool as the tungsten is withdrawn slightly, Figure 13-12.
7. When it becomes necessary to stop, taper the size of the weld pool down or leave the filler rod stuck in the weld crater.
8. Continue the root pass around the pipe. Clean and inspect it for flaws.

Procedure Open Root Pass 6G Position

Use the same setup and techniques for this weld as you have in prior procedures.

Procedure Filler and Cover Passes

For these welds you will be using a pipe joint that has the root pass already completed. The techniques described in this procedure can be applied to pipe in the other positions.

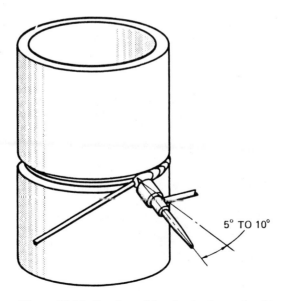

Figure 13-11 Torch position for horizontal weld

Pipe Welding Techniques

Figure 13-12 Add the filler along the top edge of the weld pool

1. After the root pass is completed the cup can no longer be supported against the pipe. You must learn to support your hand against the pipe so that you are stable but still free to move.
   Note: You can practice this skill be making stringer beads around a pipe, Figure 13-13.
2. Keep the weld deposit small and uniform as the weld progresses around the pipe. It's often a temptation to make the weld larger on top because it is easier to control but this will cause uniformity problems later.
3. The weld reinforcement must be of a suitable size for the pipe size, Figure 13-14.
4. After the weld is completed inspect it for uniformity and any flaws. Cut out the specimens, prepare them for testing, and bend-test them. If any specimens fail, continue making the weld until all specimens are of sufficient quality to pass the test.

Figure 13-13 Beading on pipe

Gas Tungsten Arc Welding of Pipe

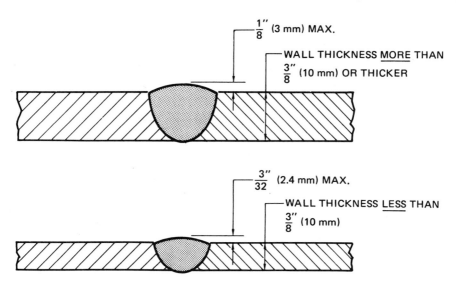

Figure 13-14 Excessively large weld beads prevent the pipe from uniformly expanding under pressure. This restriction from expanding can cause a failure to occur in or near the weld. (From Jeffus & Johnson, *Welding Principles and Applications.* Copyright 1984 by Delmar Publishers Inc.)

REVIEW

Select the letter preceding the best answer.

1. The tungsten can be moved upward along the root smoothly by
    a. resting your hand on the pipe.
    b. rocking the cup in the V-groove.
    c. turning the pipe in a set of rollers.
    d. using a little oil on the cup.

2. GTA welding is used for pipe because
    a. it's clean (there is no slag to clean up).
    b. most metals can be joined.
    c. high quality welds can be produced.
    d. All of the above.

3. A backing gas is used to
    a. keep the root surface from oxidizing.
    b. support the root pass.
    c. prevent excessive root penetration.
    d. reduce the heat needed to make the weld.

Pipe Welding Techniques

4. Where is the filler rod added to the molten weld pool in the 2G position?
    a. At the leading edge
    b. At the top edge
    c. In the center
    d. On the back top edge

5. How does a consumable insert differ from a backing ring?
    a. The backing ring does not restrict flow.
    b. The consumable insert does not melt completely.
    c. The backing ring does not act as filler metal.
    d. They both serve the same purpose and are interchangeable terms.

Unit 14

GAS METAL ARC AND FLUX-CORED ARC WELDING OF PIPE

The gas metal arc welding (GMAW) process is used for filler and cover passes. Because of a history of incomplete root fusion it has limited use today for the root pass; however, flux-cored arc welding does not have these problems. With the newer, smaller sizes of FCA welding wires FCAW is increasing in its usage. The most popular application for GMA welding on pipe is in the 1G position. Both GMAW and FCAW can be used in connection with some machine, automated, or robotic welding equipment, Figure 14-1.

MATERIALS

- One or more pieces of 4-inch to 8-inch pipe
- A GMA welding setup
- Hand grinder

Figure 14-1 This robot is making the tube to tube-sheet welds as used on a heat exchanger.

119

Pipe Welding Techniques

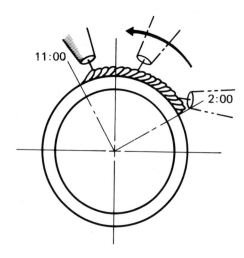

Figure 14-2 Changing gun angle for vertical-up weld

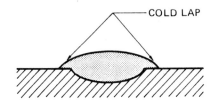

Figure 14-3 Cold lap along the edges of a weld is a common GMA problem

Procedure Stringer Beads 1G

1. Place a piece of pipe on a table about chest high.
2. Set the GMA welder to the short-circuiting metal transfer mode.
3. Start a weld at the 2:00 o'clock position and weld upward to the 11:00 o'clock position, Figure 14-2.
4. Stop and examine the weld for uniformity and quality.
   Note: The most common flaw with this weld is a lack of fusion along the toe of the weld, Figure 14-3.
5. Rotate the pipe and continue the weld around the pipe.
6. Examine the weld for uniformity and quality. If the weld does not pass continue practicing until it is mastered.

Procedure Filler and Cover Passes 1G

1. Place a properly tack welded V-grooved pipe joint, with or without the root pass, on a table about chest high.
2. Set the GMA welder to the short-circuiting metal transfer mode.
3. Start a weld at the 2:00 o'clock position and weld upward to the 11:00 o'clock position.
4. Stop and examine the weld for uniformity and quality.
5. Rotate the pipe and continue the weld around the pipe.
6. Examine the weld for uniformity and quality.
7. Complete the filler and cover passes on the pipe joint and inspect each pass until the weld is completed. Cut the specimens, prepare them for testing, and bend-test them. If any specimens fail, continue to make the weld until all the specimens pass.

Procedure Filler and Cover Passes 2G

1. Place a properly tack welded V-grooved pipe joint, with or without the root pass, on a table about chest high.
2. Set the GMA welder to the short-circuiting metal transfer mode.
3. Start the weld at a point away from you and weld using a trailing angle toward yourself, Figure 14-4.
   Note: Keep the weld size small so it's easier to control.
4. Stop and examine the weld for uniformity and quality.
5. Rotate the pipe and continue the weld around the pipe.
6. Examine the weld for uniformity and quality.
7. Complete the filler and cover passes on the pipe joint and inspect each pass until the weld is completed. Cut the specimens, prepare them for testing, and bend-test them. If any specimens fail, continue to make the weld until all the specimens pass.

Procedure Filler and Cover Passes 5G

1. Place a properly tack welded V-grooved pipe joint, with or without the root pass, on a table about chest high.
2. Set the GMA welder to the short-circuiting metal transfer mode.
3. Start a weld at the 6:30 position and weld upward to the 11:30 position, Figure 14-5.

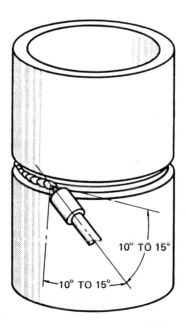

Figure 14-4 GMA gun angle for a horizontal weld

Pipe Welding Techniques

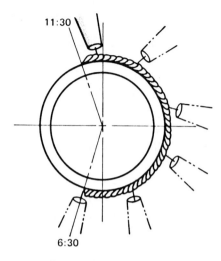

Figure 14-5 Vertical up

4. Stop and examine the weld for uniformity and quality.
5. Start a weld at the 5:30 position and weld upward to the 12:30 position.
6. Examine the weld for uniformity and quality.
7. Complete the filler and cover passes on the pipe joint and inspect each pass until the weld is completed. Cut the specimens, prepare them for testing, and bend-test them. If any specimens fail, continue to make the weld until all the specimens pass.

MATERIALS

- One or more pieces of 8-inch or larger pipe
- An FCA welding setup
- Hand grinder

Procedure Stringer Beads 1G

1. Place a piece of pipe on a table about chest high.
2. Set the FCA welder.
3. Start a weld at the 2:00 o'clock position and weld upward to the 11:00 o'clock position, Figure 14-5.
4. Stop and examine the weld for uniformity and quality.
5. Rotate the pipe and continue the weld around the pipe.
6. Examine the weld for uniformity and quality. If the weld does not pass, continue practicing until it is mastered.

Gas Metal Arc and Flux-cored Arc Welding of Pipe

Procedure Filler and Cover Passes 1G

1. Place a properly tack welded V-grooved pipe joint, without the root pass, on a table about chest high.
2. Set the FCA welder.
3. Start a weld at the 2:00 o'clock position and weld upward to the 11:00 o'clock position.
4. Stop and examine the weld for uniformity and quality.
5. Rotate the pipe and continue the weld around the pipe.
6. Examine the weld for uniformity and quality.
7. Complete the filler and cover passes on the pipe joint and inspect each pass until the weld is completed. Cut the specimens, prepare them for testing, and bend-test them. If any specimens fail, continue to make the weld until all the specimens pass.

Procedure Filler and Cover Passes 2G

1. Place a properly tack welded V-grooved pipe joint, without the root pass, on a table about chest high.
2. Set the FCA welder.
3. Start the weld at a point away from you and weld using a trailing angle toward yourself, Figure 14-4.
   Note: Keep the weld size small so it's easier to control.
4. Stop and examine the weld for uniformity and quality.
5. Rotate the pipe and continue the weld around the pipe.
6. Examine the weld for uniformity and quality.
7. Complete the filler and cover passes on the pipe joint and inspect each pass until the weld is completed. Cut the specimens, prepare them for testing, and bend-test them. If any specimens fail, continue to make the weld until all the specimens pass.

Procedure Filler and Cover Passes 5G

1. Place a properly tack welded V-grooved pipe joint, without the root pass, on a table about chest high.
2. Set the FCA welder.
3. Start a weld at the 6:30 position and weld upward to the 11:30 position.
4. Stop and examine the weld for uniformity and quality.
5. Start a weld at the 5:30 position and weld upward to the 12:30 position.
6. Examine the weld for uniformity and quality.
7. Complete the filler and cover passes on the pipe joint and inspect each pass until the weld is completed. Cut the specimens, prepare them for testing, and bend-test them. If any specimens fail, continue to make the weld until all the specimens pass.

Pipe Welding Techniques

REVIEW

Select the letter preceding the best answer.

1. Which method of metal transfer is the most frequently used on pipe?
   a. Short-circuiting
   b. Globular
   c. Spray arc
   d. Pulsed arc

2. Which position(s) is/are most GMAW pipe welds made in?
   a. 1G
   b. 1G and 5G
   c. 2G
   d. All but 6G

3. Why is the GMAW process not used for the root pass?
   a. It penetrates too much.
   b. Short wire sections (whiskers) often are left stuck inside the pipe.
   c. There is often little root fusion.
   d. The weld pool is too large to control.

4. What is the most common beading problem with GMAW?
   a. Too much penetration
   b. Cold lap at the toe
   c. Too large a build-up
   d. Poor gas coverage of the root

5. GMA welding of pipe is often done in conjunction with
   a. machine welding.
   b. automated equipment.
   c. robotic equipment.
   d. All of the above.

6. What is the major advantage of using FCA welding of pipe as compared to using GMA welding?
   a. FCAW has a faster filling characteristic.
   b. FCAW is better in the vertical-down position.
   c. FCAW has deeper penetration.
   d. FCAW has less spatter.

7. Why should the FCA weld bead be kept small when welding out of position?
   a. It's easier to control the bead shape.
   b. The molten weld pool is easier to see.
   c. There will be less flux trapped.
   d. Both b and c.

8. Why can the root pass be welded with FCAW?
   a. The visibility is better.
   b. There is a flux to protect the root side of the joint.
   c. The bead shape can be easily controlled.
   d. The root fusion is complete.

Unit 15

PIPE FITTING

It is beyond the scope of this text to give detailed consideration to templet making, layout, and fitup of welded-pipe connections. However, if students of pipe welding are to gain experience in making lateral, T- and Y-branch connections, practice materials must be available.

To gain experience in fabrication and avoid unnecessary expense, welding students should make their own fittings, cutting them from available pipe stock.

The simplest method of making a T connection is to obtain a templet made by an expert templet maker, apply it to the pipe to be cut, and mark the outline with soapstone. Centerpunch and flame-cut the pipe to the desired contour as indicated in Figures 15-1 through 15-4. Observe that the centerline of the cutting tip is held square with the pipe throughout the entire cut.

For accuracy of fit, the templet maker must take into consideration the thickness of the pipe to be cut. The inside of the branch connection will then fit outside of the *header,* which is a main pipeline to which the branches are connected. Figure 15-5 shows a branch fitted to the header. Observe the natural V which this type of joint provides for welding.

If, instead, the outside of the branch is laid out to fit the header, the angle of the cutting tip must be constantly changed to ensure proper fitup as indicated in Figure 15-6.

Pipe can be laid out by use of mechanical devices such as the *contour marker* which is being used to lay out a lateral on 4-inch pipe in Figure 15-7. These markers are highly accu-

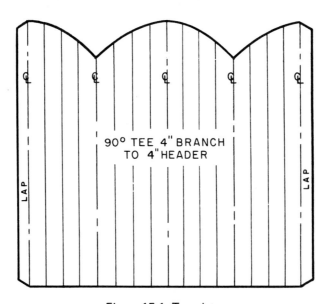

Figure 15-1 Templet

Pipe Fitting

Figure 15-2 Using a templet to mark the pipe with soapstone

Figure 15-3 Centerpunching along the marking

Pipe Welding Techniques

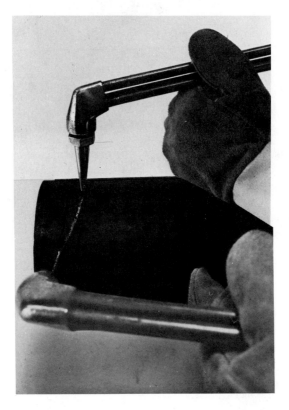

Figure 15-4 Cutting the contour of the joint

rate when properly used. They require a certain amount of practical use to gain experience. Once familiar with this tool, the user can lay out an almost infinite variety of Ys, Ts, laterals, and miters within the limits of the particular tool being used. Figure 15-8 shows the excellent fitup of the lateral laid out in figure 15-7.

Note: In this type of layout, the branch is cut into the header. If properly made, both the bore of the pipe and the outside diameter fit perfectly.

Pipe Fitting

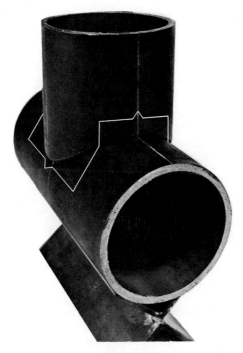

Figure 15-5 Inside of branch fitting outside of header

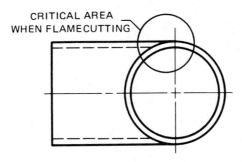

Figure 15-6 Outside of branch laid out to fit outside of header

129

Pipe Welding Techniques

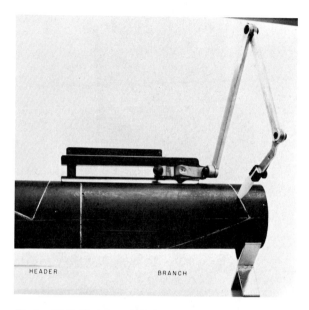

Figure 15-7 Pipe lateral being laid out with contour marker

Figure 15-8 Fitup of lateral cut from contour layout

REVIEW

A. Select the letter preceding the best answer.

1. In a lateral-branch connection of welded pipe,
   a. the layout can be made only with a contour marker.
   b. the branch pipe forms a leg of a 90-degree turn.
   c. the branch pipe is cut into the header.
   d. manufactured fittings to connect laterals to a piping system increase piping assembly costs.

2. The contour of welded-pipe connections made in small shops is most likely to be produced by
   a. a special cutting tip.
   b. electric arc cutting.
   c. machining.
   d. oxyacetylene cutting.

3. Fitup of pipe connections to be welded refers to
   a. making all fittings on the job.
   b. the process of assembling prepared fittings.
   c. the total process from design of contours through finished welding.
   d. cutting and setting up two pieces to be welded.

4. In a small shop, the most common method of providing for branch connections and fittings is to use
   a. ready-to-weld manufactured fittings and/or ready-made templets.
   b. shop-made fittings made from pipe stock with custom-made templets.
   c. ready-made commercial templets with mechanical markers.
   d. mechanical markers.

B. Make a cross-sectional sketch of the lap section of a 90-degree branch connection.

Unit 16

90-DEGREE BRANCH CONNECTIONS

This unit will give the prospective pipe welder experience in laying out, cutting, fitting up, and welding 90-degree branch connections on equal size pipe in various positions.

MATERIALS

- One piece, 4-inch schedule 40 pipe, 5 inches long
- One piece, 4-inch schedule 40 pipe, 9 inches long
- 1/8-inch or 5/32-inch E6010 electrode
- DC welding machine

Procedure

1. Lay out and cut the branch connection from the 5 inch long pipe, using the proper templet and checking for dimensional accuracy before making the cut, Figure 16-1.
   Note: This is the first instance in which dimensions have been applied to the symbol drawing. This is to indicate that all branch connections must maintain a high degree of dimensional accuracy if they are to serve their intended purpose.
2. Using the proper templet, lay out and cut the hole in the 9 inch long header.

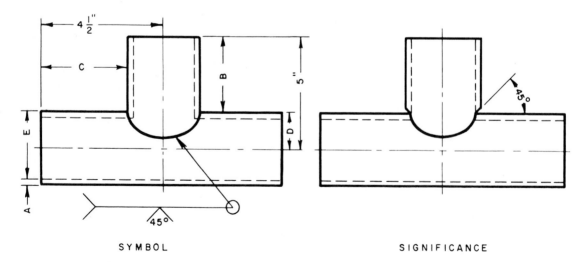

Figure 16-1 90-degree branch connection

90-degree Branch Connections

CENTERLINES LAID OUT FROM TEMPLET MAKE
FOR EASIER AND MORE ACCURATE SETUP.

Figure 16-2 Setup with centerlines aligned   Figure 16-3 Beveled branch connection

3. Clean all slag or oxide from both pieces.
4. Assemble and tack the pieces after making sure all dimensions are accurate and that the angle is exactly 90 degrees.
   Note: It will add to the speed and accuracy of the setup process if the centerlines of the branch and header are lightly centerpunched, while the wraparound templet is applied to the pipe. Draw these lines with a straightedge and soapstone. When the centerlines coincide at the setup stage of the process as shown in Figure 16-2, the two pieces of pipe are in correct relation to each other.

   When the branch connection is flame-cut to present a groove to ensure complete fusion into the inside of the branch, or whether no beveling is required, is dictated by the manufacturer's procedure specifications. In general, the student gains more practice if the branch connection is beveled as in Figure 16-3.
5. Set up the assembly at, or slightly below, eye level with the branch connection in a vertical position.
   Note: A shop weld on a fitting of this size would undoubtedly be made as a rolling

Pipe Welding Techniques

weld. However, much more experience will be gained in making this commonly welded field joint if the header is set up in a horizontal fixed position.

6. Start the root pass at the side of the header centerline at the 12:30 position as looking down from the branch, Figure 16-4. Weld across the tack and downward. Keep changing the angle of the electrode to conform to the change in contour of the weld from a fillet to a lap.

    Note: An analysis of this joint shows that the weld progresses from a flat fillet to a slightly vertical-down fillet to a vertical-down lap and, finally, to a horizontal lap weld in only a quarter of the distance around the branch. As the weld progresses through the next quarter, the order of change is reversed. The student must constantly change the electrode angle and weave pattern to produce a weld of uniform appearance.

7. When stopping to change position, clean the initial bead and inspect both the face of the weld and the root as done previously in other units. When continuing this bead, correct for any defects observed in the initial inspection.

8. After the completed root pass has been cleaned and inspected, make the filler pass, using a slight weave motion.

9. After cleaning and inspecting the filler pass, make the final pass, using much the same technique.

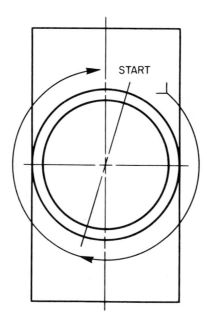

Figure 16-4 Run the bead starting at 12:30

90-degree Branch Connections

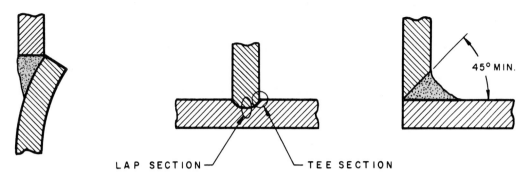

Figure 16-5 Correct cross section through lap and tee of 90-degree branch

10. Clean and inspect this bead. Notice the contour of the finished weld. Since any abrupt change in contour tends to concentrate stresses in that area, the welder should make a smooth transition from one pass to the next. Figure 16-5 indicates the acceptable contour of the welds in T joints.
11. After sufficient practice in making T joints in the above position, prepare and set up more identical joints, but position the branch so that it lies in a horizontal position. Almost half of the welding is done in the overhead position, and the rest is divided between vertical and flat welding.
    Note: As in previous welds, the direction of welding will be dictated by the manufacturer's or contractor's procedure specifications. It will be well to make some of these joints using the vertical-down procedure and others using the vertical-up procedure in order to gain skill, experience, and speed in both.
12. Make an entire joint using the vertical-down method. After the joint has been completed and visually inspected, cut a specimen from the fillet portion of the joint and one from the lap portion.
13. Grind, polish, and macroetch these specimens and thoroughly inspect for flaws such as lack of root fusion, slag inclusions, gas pockets, and undercutting.
14. Make more branch connections, using the vertical-up method until this technique has also been mastered.
15. Set up more 90-degree branch connections with the branch pointing directly down. Observe that with the exception of a small amount of the weld on the sides, which is considered a horizontal lap weld, the remainder is overhead, changing from overhead lap to overhead fillet as the welding progresses.
16. Make a complete joint in this position.

Pipe Welding Techniques

REVIEW

1. Refer to Appendix I and the illustration below. Give the dimensions of A, B, C, and D.

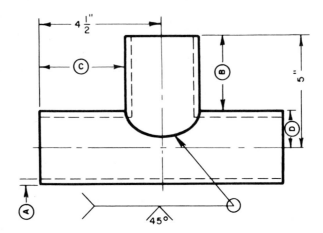

2. If the pipe were double-extra strong, what would the dimensions be for A, B, C, and D in the connection shown above?

3. Arrange the following operations in making the 90-degree branch connection in the best performance sequence by placing the letters in the correct order.

   To set up a vertical branch connection on a fixed header
   a. tack the vertical branch to the header.
   b. check the angle between the branch and the header with a square, being sure it is vertical and on the centerline of the header.
   c. weld the joint.
   d. check the side of the branch with a level to make sure it is plumb on this location.

Select the letter preceding the best answer.

4. In making the weld for this assembly with the branch pointing up,
   a. use a weave motion in making the filler pass.
   b. position the center of the setup above eye level.
   c. use a rolling weld to get more practice.
   d. keep the electrode angle constant.

5. When tacking fittings or pipe for accurate alignment,
   a. first run a light bead all around the joint.
   b. place four tacks in sequence all around the joint.
   c. place one tack, and then another 180 degrees away. Repeat with equal spacing for the other two tacks.
   d. use as few tacks as possible.

Unit 17

LATERAL PIPE CONNECTIONS

This unit is designed to give the pipe welding student experience in the layout, flame-cutting, fitup, and welding of lateral pipe connections in the most frequently used positions.

MATERIALS

- One piece, 4-inch schedule 40 pipe, 7 inches long
- One piece, 4-inch schedule 40 pipe, 10 inches long
- 1/8-inch or 5/32-inch E6010 and E6011 electrodes
- Welding machine

Procedure

1. Lay out and cut the branch connection from the 7-inch pipe, using the proper wrap-around templet. Mark the centerline to facilitate the setup.
2. Lay out and cut the hole in the header. Check the dimensions with Figure 17-1.
3. Thoroughly clean both pieces, set the branch on the header with the centerline properly aligned, and examine the setup to determine those areas where beveling with a handheld cutting torch would provide for complete root fusion.

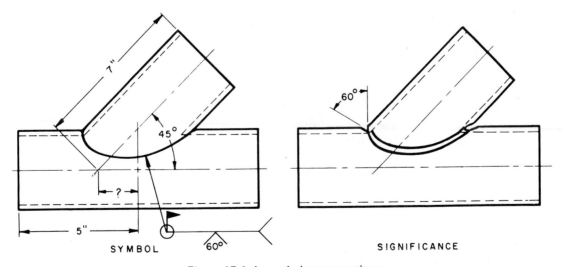

Figure 17-1 Lateral pipe connections

Pipe Welding Techniques

4. Mark these areas and flame-cut the bevels.

    Note: Beveling in the area of the 135-degree intersection of the branch and header can be carried out with relative ease. As the line of intersection of the two pieces approaches the area of the 45-degree intersection, proper beveling becomes more difficult and finally impossible by the standard procedure. Many laterals are made using no bevel in the area of the 45-degree intersection. In this case an extra-heavy fillet weld is made to partially compensate for the lack of fusion in the root. However, every effort should be made to obtain as complete penetration and fusion as possible. Engineering data indicates that the allowable working stress for laterals is only 40 percent of that of the pipe from which it is cut. Figure 17-2 shows a lateral beveled all-around and ready for welding.

5. Set up the assembly at eye level or slightly below with the header essentially horizontal and the branch in a vertical plane.

6. With the machine set for normal amperage, or slightly higher, strike an arc and start the root pass. Do not attempt to start the bead at the 45-degree intersection. Also, plan the procedure in such a manner that it will not be necessary to stop any bead at or near the 45-degree intersection. This is the area in which most difficulties will be encountered. If the procedure is planned so that the arc is in continuous operation at the 45-degree

Figure 17-2 Completely beveled lateral

Lateral Pipe Connections

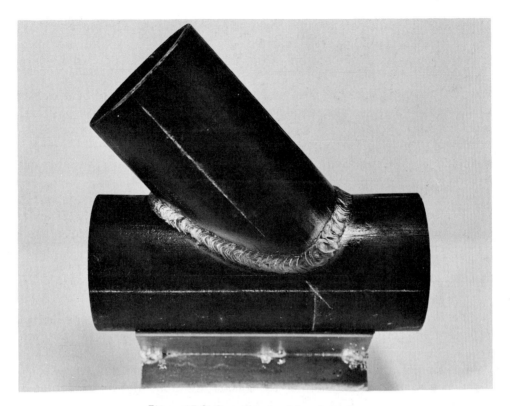

Figure 17-3 Completed weld on lateral

    intersection and for about 2 inches on each side of it, the probability of lack of fusion, slag inclusions, undercutting, or excessive buildup is minimized.
7. Clean and inspect both the face and root of this initial bead. Be especially critical of those areas in which difficulties were encountered.
8. Make the filler bead using slight weaving motion as in previous units. Be sure to vary the weave pattern to compensate for the changing type and position of the joint.
9. After this bead has been inspected, make a fillet weld starting about 20 degrees off center from the 45-degree intersection and continue it across the intersection for another 20 degrees. This provides additional reinforcement in this highly stressed area when the final pass is made. It also allows the final pass to be made with a smoother continuity than would be possible if the reinforcement were applied after the final pass had been completed. Figure 17-3 shows the completed bead.
10. Make the final pass. Figure 17-3 shows the completed weld.
11. Clean and inspect the completed joint.
12. Cut a section from the 45-degree area and macroetch. Examine this specimen for evidence of lack of fusion, holes, and undercut.
13. Set up an additional lateral with the branch and header both in a horizontal plane.

Pipe Welding Techniques

14. Before welding, anticipate the positions and types of welds to be made so that unnecessary pauses during welding are eliminated.
15. Start the root pass on the bottom of the joint and weld in an upward direction around the joint.
    Note: If the procedure has been carefully planned, it should not be necessary to stop at the 45-degree intersection to change either the electrode or position.
16. Make the filler and final passes using the techniques learned in previous units. Be sure to observe all previous precautions as to procedure and inspection.
17. Make additional laterals with the branch pointing downward. In this setup most of the welding is in the overhead position.
    Note: In this position, the branch will be in the way to such an extent that the proper electrode angle and continuity of the weld will be extremely difficult to maintain in the 45-degree area. In actual practice, this position of a lateral would only be used to tie into an existing pipe system. Even then, if shutdown time permitted, it would be more practical to cut a section from the existing pipe system, then fit a previously made lateral into the space and weld the all-around seams.
18. Clean and inspect this completed joint as in previous steps, noting any faults that may be present and correcting them when making additional laterals in this position.

Note: At this point, the pipe welder should have gained enough skill, knowledge, and experience to meet all the normal and special situations that are encountered in welding a piping system. Pipes do not always lie in a horizontal or vertical plane so some welds must be made at an intermediate angle. A close examination and appraisal of any of these joints will show that the welding procedure, positions, and techniques are very similar to those which have already been learned.

REVIEW

Select the letter preceding the best answer.

1. To avoid difficult beveling in the area of the 45-degree intersection,
    a. use a cutting torch for beveling this area.
    b. substitute a heavy fillet weld for the bevel.
    c. start the bead in this area.
    d. use stringer beads in this area.

2. The 45-degree area is difficult to prepare and weld because of the
    a. restricted angle in which to work.
    b. difficulty of making an accurate setup.
    c. excessive root fusion.
    d. difficulty of keeping the branch and header aligned.

Lateral Pipe Connections

3. The highest stress concentration in this welded joint under operating pressures would be
   a. at the section opposite the 45-degree area.
   b. in the 45-degree area.
   c. uniform throughout the joint.
   d. at the side of the branch.

4. The procedure for making the root pass should be planned so that
   a. the bead will not be started near the 45-degree area.
   b. the bead will be stopped at the 45-degree intersection.
   c. the arc is continuous except at the 45-degree intersection.
   d. the bead can be stopped or started anywhere around the joint.

5. What would be the most effective and efficient position for welding a lateral fitting?
   a. Place the header in a vertical position with the lateral pointing down.
   b. Place the header in a vertical position with the lateral pointing up.
   c. Arrange it so that the whole fitting is rigidly clamped in place.
   d. Put in a fixture that permits rotation of the fitting.

Appendix I

SIZES AND TYPES OF COMMERCIAL PIPE

Nominal Pipe Size	Outside Diameter	Nominal Wall Thickness			
		Standard	Schedule 40	Extra Strong	XX Strong
1/8	0.405	0.068	0.068	0.095	—
1/4	0.540	0.088	0.088	0.119	—
3/8	0.675	0.091	0.091	0.126	—
1/2	0.840	0.109	0.109	0.147	0.294
3/4	1.050	0.113	0.113	0.154	0.308
1	1.315	0.133	0.133	0.179	0.358
1 1/4	1.660	0.140	0.140	0.191	0.382
1 1/2	1.900	0.145	0.145	0.200	0.400
2	2.375	0.154	0.154	0.218	0.436
2 1/2	2.875	0.203	0.203	0.276	0.552
3	3.5	0.216	0.216	0.300	0.600
3 1/2	4.0	0.226	0.226	0.318	—
4	4.5	0.237	0.237	0.337	0.674
5	5.563	0.258	0.258	0.375	0.750
6	6.625	0.280	0.280	0.432	0.864
8	8.625	0.322	0.322	0.500	0.875
10	10.75	0.365	0.365	0.500	—
12	12.75	0.375	0.406	0.500	—
14 O.D.	14.0	0.375	0.438	0.500	—
16 O.D.	16.0	0.375	0.500	0.500	—
18 O.D.	18.0	0.375	0.562	0.500	—
20 O.D.	20.0	0.375	0.593	0.500	—
22 O.D.	22.0	0.375	—	0.500	—
24 O.D.	24.0	0.375	0.687	0.500	—
26 O.D.	26.0	0.375	—	0.500	—
30 O.D.	30.0	0.375	—	0.500	—
34 O.D.	34.0	0.375	—	0.500	—
36 O.D.	36.0	0.375	—	0.500	—
42 O.D.	42.0	0.375	—	0.500	—

All dimensions are given in inches. The decimal thickness listed for the respective pipe sizes represent their nominal or average wall dimensions. The actual thicknesses may be as much as 12.5% under the nominal thickness because of mill tolerance.

The table lists the pipe sizes and wall thicknesses established as: standard weight, extra strong, and double extra strong.

Appendix II

CONVERSION FACTORS:
U.S. CUSTOMARY UNITS AND SI METRIC UNITS

TEMPERATURE

°F to °C	°F [−] 32 [=] _____ [X] .555 [=] _____ °C
°C to °F	°C [X] 1.8 [=] _____ [+] 32 [=] _____ °F
°C to K	°C [+] 273.1 [=] _____ K
K to °C	K [−] 273.1 [=] _____ °C

LINEAR MEASUREMENT

in to mm	in [X] 25.4 [=] _____ mm
mm to in	mm [X] .0394 [=] _____ in
ft to mm	ft [X] 304.8 [=] _____ mm
mm to ft	mm [X] .00328 [=] _____ ft
ft to m	ft [X] .3048 [=] _____ m
m to ft	m [X] 3.28 [=] _____ ft

AREA MEASUREMENT

in² to mm²	in² [X] 645.2 [=] _____ mm²
mm² to in²	mm² [X] .00155 [=] _____ in²

WEIGHT

lb to N	lb [X] 4.448 [=] _____ N
N to lb	N [X] .2248 [=] _____ lb
lb to kg	lb [X] .4536 [=] _____ kg
kg to lb	kg [X] 2.205 [=] _____ lb

PRESSURE

psi to kg/mm²	psi [X] .000703 [=] _____ kg/mm²
kg/mm² to psi	kg/mm² [X] 6894.7 [=] _____ psi

Pipe Welding Techniques

VELOCITY

in/min to mm/sec	in/min ☒ .4233 ☐ _____ mm/sec
mm/sec to in/min	mm/sec ☒ 2.362 ☐ _____ in/sec
cfh to L/min	cfh ☒ .4719 ☐ _____ L/min
L/min to cfh*	L/min ☒ 2.119 ☐ _____ cfh

*cfh = ft^3/h

(From Jeffus & Johnson, *Welding Principles and Applications.* Copyright 1984 by Delmar Publishers Inc.)

Appendix III

WELDING CODES AND SPECIFICATIONS

A *welding code* is a detailed listing of the rules or principles which are to be applied to a specific classification or type of product.

A *welding specification* is a detailed statement of the legal requirements for a specific classification or type of product. Products manufactured to code or specification requirements commonly must be inspected and tested to assure compliance.

There are a number of agencies and organizations that publish welding codes and specifications. The application of the particular code or specification to a weldment can be the result of one or more of the following requirements:

- Local, state, or federal government regulations
- Bonding or insuring company
- End user (customer) requirements
- Standard industrial practices

The three most popular codes are:

#1044, American Petroleum Institute—used for pipelines

Section IX, American Soceity of Mechanical Engineers—used for pressure vessels

D1.1, American Welding Society—used for bridges and buildings

The following organizations publish welding codes and/or specifications.

AASHT
    American Association of State Highway and Transportation Officials
    444 North Capitol Street, NW
    Washington, DC 20001

AIAA
    Aerospace Industries Association of America
    1725 DeSales Street, NW
    Washington, DC 20036

AISC
    American Institute of Steel Construction
    400 North Michigan Avenue
    Chicago, Illinois 60611

ANSI
American National Standards Institute
1430 Broadway
New York, New York 10018

API
American Petroleum Institute
2101 L Street, NW
Washington, DC 20037

AREA
American Railway Engineering Association
Suite 403
2000 L Street, NW
Washington, DC 20036

ASME
American Society of Mechanical Engineers
345 East 47th Street
New York, New York 10017

AWWA
American Water Works Association
6666 West Quincy Avenue
Denver, Colorado 80235

AWS
American Welding Society
550 NW LeJeune Road
Miami, Florida 33126

AAR
Association of American Railroads
1920 L Street, NW
Washington, DC 20036

MIL
Department of Defense
Washington, DC 20301

SAE
Society of Automotive Engineers
400 Commonwealth Drive
Warrendale, Pennsylvania 15096

(From Jeffus & Johnson, *Welding Principles and Applications.* Copyright 1984 by Delmar Publishers Inc.)

INDEX

Argon, use of, 115

Beads. *See also* Flux-cored arc welding; Gas-metal arc welding; Gas-tungsten arc welding; Horizontal pipe; Lateral pipe connections; Qualifying test
  arc strikes outside weld joint, 69
  bead size, taper of, 71
  bracing of self for making, 67
  electrode, horizontal, 72
  electrode, vertical, 73
  fixed-down, 5G, 72
  fixed-up 5G, 70–71
  horizontal butt welds, 103
  horizontal rolled procedures, 1G, 67–68, 70
  low-hydrogen, 58
  materials, 67
  pipe position, 68
  stringer beads, 69
  turning of pipe, 70
  vertical down, 51
  vertical 2G, 72–74
  view of weld, 73
  weave, 101
Beveling
  air-powered tool, 76
  by hand, 80
  by machine, 80
  overhead single V-groove, 64
  plate, 55, 59
  qualifying test, 26
Butt joint fitting, 76–85
  air-powered pipe beveling tool, 76
  automatic cutting machine, 77
  butt joint assembly, 81, 83–84
  coin for root spacing, 84
  correct cut, 81
  cutting pipe, 79
  devices for alignment, 83
  double torch holder, 77
  hand beveling of pipe, procedure for, 80
  hot tack weld, alignment of joint, 84
  locations to avoid for tack welds, 85
  machine beveling of pipe, procedure, 80
  materials, 78
  oxyfuel torch for beveling, 79
  pipe alignment clamps, 82
  pipe joint face, 80
  square butt joint fitting, 81
  tack welding pipe joint, 84
  turntable, 78
  typical joint design, 82

Clamps, for pipe, 16, 82
Code welding, 7–15
  difficult positions, ASME code on, 14–15
  procedure qualifications, 7, 9
  test positions, 13
  welder performance qualifications, 9–12
  welding procedure specification form, 8
Codes and specifications for welding, organizations, 5–6, 147–48
Conversion factors, numerical, 145–46

Defects in welds, causes, 88
Dry well assembly, welding of, 5

Electrodes
  angles, 90
  E6010, 32, 35, 37, 39–41, 43, 50–52, 55–57, 58, 61–63, 64, 67–68, 70, 72–73, 87, 91–93, 99–104, 107–08
  E6011, 32, 35, 37, 39–41, 43, 50–52, 55–57, 58, 61–63, 64, 67–68, 70, 72–73, 87, 91–93, 99–104
  E7018, 35, 38–39, 41–42, 43–44, 57–58, 58–60, 63, 68–70, 71, 73, 89–90, 93–94, 94–96, 104, 107–08
  horizontal welds, 72, 103
  key hole, 36
  low-hydrogen, 58, 63
  open-root pass, vertical-down, 5G, 94–96
  overhead single V-groove, 63
  qualifying test, 25
  stepping motion, 87
  types, 25
  vertical welds, 73

Fillet welds
  in pipe, 13
  in plate, 13
Flux-cored arc welding of pipe, 124–25
  filler and cover pass
    5G, 125
    1G, 125
    2G, 125
  materials, 124
  stringer beads, 1G, 124
  45°-inclined welds
    materials, 107
    restricting ring, 107, 108
    6G open root, 107–08
    6GR open root, 108
    test positions, 107

Gas metal arc welding, 121–24
  cold lap, problems with, 122

147

Pipe Welding Techniques

Gas metal arc welding *'Continued'*
  filler and cover passes
    5G, 123-24
    1G, 122
    2G, 123
  gun angle, changing of, 122
  horizontal weld, angle for gun, 123
  materials, 121
  robots in, 121
  stringer beads, 1G, 122
  vertical-up, 124
Gas tungsten arc welding, 111-19
  automatic unit for large-diameter pipe, 111
  backing rings, specialized, 114
  centerline cracking of bead, 113
  cup, support in groove, 112
  5G, 116-18
  1G, 112-16
  open root pass, 1G position, 112-14
  reasons for, 111
  rods, support for, 112
  torch motion, 113
Groove welds
  in pipe, 13
  in plate, 13

Horizontal pipe, 1G and 5G positions, welding of
  back-grinding to remove discontinuities, 89
  bead width, 92
  butt welds, fixed position, 91
  defects in pipe welds, causes, 88
  electrode angles, 90
  electrode, stepping motion for, 87
  end points, welding passes, 92
  filler and cover pass, 89-90
  5G, 90-96
  grinding stone, holding of, 88
  materials, 86
  multiple-pass welds, staggering of endpoints, 93
  1G, 86-90
  open root pass, 87, 89
  open root pass vertical-down 5G, 94-97
    concave beads, problems with, 96
    cross-section, 96
    electrode angle, 96, 97
    electrode position, 95
    molten pool, shape of, 95
    procedure, 94-96
  pipe for, 86
  rocking motion for electrode, 94
  rollers, 86
  rolling pipe joint, acceptable example, 91
  root pass cleanup, 88
  side fusion of weld, 94
  test specimen, removal order, 94
  vertical-up 5G open root pass, 91-94
Horizontal 2G, 41-42

Lateral pipe connections
  beveling, 140
  completed weld on lateral, 141
  diagram, 139
  downward branches, 142
  fillet weld, 141
  45°-lateral, bead for, 140-41
  materials, 139
  procedure, 139-42
Leather jacket for welding, 42

Macroetching, 27-29

90°-branch connections, 134-37
  bead running, 136
  beveled branch connections, 135
  centerline alignment, 135
  correct cross-sections, 137
  diagram, 134
  line drawing, 135
  materials, 134
  procedure, 134-37
  root pass, 136-37
  T-joints, 137

Open roots
  internal rings, 116-17
  overhead single V-groove, 64
  pass, 87, 89
  vertical-down, 5G, 94-97
  vertical-up, 5G, 91-94
Oxide scale, 26
Overhead 4G, 42-44
Overhead single V-groove welds 4G, 61-64
  backing strip, 61-63
  beads, 62
  bevels, 64
  diagram, 61
  hand and electrode positions for, 63
  low-hydrogen electrode, 63
  nick-break test specimen, 64
  open root, 64
  slag, 62-63
  test specimens, 62-63
  undercutting, 62
  without backing strip, 64
Oxide scale, 26

Pipe
  methods of joining, 2
  production of, 1
  sizes and types of, commercial, 144
  uses, 1
  welded systems, requirements of
    field welding, 4
    flow comparison, 4
    physical demands, 2-3
    pipe fittings, 2
    thickness of joint, 3
    threaded joint, 3

Qualifying test, procedure aids for

Index

backing, 26
base metal specimen bend, 24
beads, 27
beveling, 26
direction of rolling, 25-26
electrodes, 25
groove weld, 26
joint preparation, 26
low-ductility failure, 25
multiple-pass weld, 27
oxide scale, 26
passes, number of, 27

Rings, internal
 backing, 114
 chill, 114
 consumable insert, 114-15
 excess-size beads, problems with, 119
 filler and cover passes, 117-19
  addition of, 118
  beading on pipe, 118
 joint with consumable, 116
 open root pass
  5G position, 116
  6G position, 117
  2G position, 117
 position of rings, 115
 root pass with insert, 1G position, 115-16
Root pass, 32-44. *See also* Open roots
 backing strip, 35, 37-39, 40-41, 43-44
 edge preparation, 32
 electrode movement for key hole, 36
 grinding back of root pass, 37
 horizontal, 41
 incomplete root fusion, 38
 materials, 32
 open root, 32, 35, 39, 41, 43
 overhead, 42-44
 vertical-down, 39-41
 vertical root with plates, 33-34, 38

Soapstone, 129

T connections, making of, 128
Tack welds, 84-85
Templets, 128-32
 centerpunching, 128, 129
 contour marker, 128
 header, 128
 inside of branch fitting, outside of header, 131
 joint, cutting contour of, 130
 lateral cut, fit-up of, 132
 and marks on pipe, 129
 outside of branch, 131
 pipe lateral layout, 132
Testing methods, 15-22
 destructive, 17-18
 examples, 20

face bend specimens, 18
guided bend jig, 19, 21
guided bend test, 18-20
penetrant, 15, 17
proof, 15
radiographic, 16
reduced section tension test, 20-22
 specimen preparation for, 22
root bend specimens, 18
specimen preparation, 18
specimen removal locations, 19
tool mark direction, 19
ultrasonic, 15, 17
visual, 15
Thermocouples
 clamps, 27
 insulation, 26
 leads, 26
 spot welding, 26

V-groove, single, horizontal welds 2G
 beveled plate, 55, 59
 low-hydrogen bead sequence, 58
 passes, 56-57, 59
 with backing strip, 55, 57-58
 without backing strip, 58, 60
V-groove, single, vertical welds 3G
 bevel setup, 52
 materials, 48
 open root, 49-50
 reverse side, vertical-down root pass, 52
 undercut second bead, 49
 vertical beveled butt joint, with backing strip, 48
 vertical-down, 50-51
 vertical-down beads, 51
 vertical-up, 47
 without backing strip, 51-52
Vertical down 3G, 39-41
Vertical pipe, 2G position welding, 99-105
 arc position, 100
 bead-making order, horizontal butt welds, 103
 cleaning of welds, equipment for, 104
 current control, 100
 filler pass, weave motion, 101
 final pass as weave bead, 101
 horizontal butt welds, vertical pipe, 99
 horizontal welds, electrode angles for, 103
 horizontal welds, multipass bead positions for, 105
 materials, 99
 open root, 99-102
 open root, E7018 electrodes, 104
 open root, and stringer beads, 102-04
 sag, 100
 test specimen removal order, 102
 undercut, 100
 weld pool contact with solid metal, 100
Vertical up 3G, 32, 35, 37-39

149